FIREFIGHTER EXAM PREP

The Definitive Study Guide to Pass the Exam on Your First Attempt with Confidence

Part A: Preface

Introduction

Embarking on the path to becoming a firefighter in the United States is a commendable and ambitious goal. This profession demands not only physical prowess but also a sharp mind, emotional resilience, and a profound commitment to serving the community. The initial step in this journey involves clearing the Firefighter Entrance Exam, a hurdle designed to assess a candidate's readiness for the challenges and responsibilities of firefighting. This exam is multifaceted, covering a range of topics from basic mathematics and reading comprehension to more specialized knowledge in fire science and emergency medical procedures. Understanding the structure and content of this exam is paramount for aspiring firefighters, as it lays the foundation for their future training and career.

The Firefighter Entrance Exam is not uniform across the board; it varies significantly from one jurisdiction to another. However, certain core elements remain consistent, such as the need for a strong grasp of physical science, mechanical reasoning, and critical thinking skills. Candidates are also tested on their physical fitness through rigorous assessments that simulate the demanding conditions of actual firefighting. Preparing for this exam requires a strategic approach, combining academic study with physical training and mental preparation.

To navigate this complex preparation process, candidates must arm themselves with the right resources and strategies. This includes familiarizing themselves with the exam's format, identifying areas for improvement, and developing a comprehensive study plan. It's also beneficial to engage in practical exercises that mimic the physical and mental challenges of firefighting. This holistic approach to exam preparation not only aids in passing the exam but also in building the foundational skills necessary for a successful career in firefighting.

Moreover, the journey to becoming a firefighter is enriched by understanding the values and principles that underpin the profession. Firefighters are not just emergency responders; they are integral members of their communities, embodying courage, dedication, and a selfless commitment to the welfare of others. As such, preparing for the Firefighter Entrance Exam is also about cultivating these qualities within oneself, ensuring that candidates are not only technically proficient but also morally aligned with the ethos of the firefighting community.

Building on the foundational knowledge and skills is just the beginning. Aspiring firefighters must also stay informed about the latest developments in fire science, emergency medical services, and rescue operations. This continuous learning process is crucial, as the field of firefighting is ever-evolving with advancements in technology and changes in fire codes and safety regulations. Engaging with the firefighting community through forums, workshops, and seminars can provide valuable insights and foster a sense of camaraderie among candidates and seasoned professionals alike.

In addition to academic and physical preparation, mental and emotional readiness cannot be overstated. The ability to remain calm under pressure, make quick decisions in emergency situations, and cope with the emotional toll of the job are essential traits for any firefighter. Techniques such as stress management exercises, meditation, and scenario-based training can be instrumental in developing these capabilities. Furthermore, mentorship from experienced firefighters can offer guidance, support, and practical advice to navigate the challenges of both the entrance exam and the profession itself.

Lastly, the importance of community involvement and volunteer work should not be overlooked. Participating in community service and volunteer firefighting can provide hands-on experience, enhance a candidate's resume, and demonstrate a genuine commitment to serving others. These experiences not only enrich a candidate's personal and professional growth but also strengthen their application and readiness for a career in firefighting.

In conclusion, preparing for the Firefighter Entrance Exam is a comprehensive endeavor that extends beyond mere academic study. It encompasses physical conditioning, mental and emotional preparation, continuous learning, community involvement, and the development of a strong moral compass. By adopting a well-rounded approach to preparation, candidates can enhance their chances of success on the exam and lay a solid foundation for a fulfilling career dedicated to safeguarding lives and property.

Chapter 1: Steps to Become a Firefighter

1. School Certificates

A high school diploma or GED is the fundamental educational requirement for aspiring firefighters, serving as a testament to one's ability to complete basic education and grasp essential knowledge. This prerequisite ensures that all candidates have a solid foundation of reading, writing, and mathematical skills necessary for the complex training and responsibilities they will face in the firefighting profession. Beyond this basic qualification, additional education such as college courses or degrees in fire science, emergency medical services, or related fields can significantly bolster a candidate's prospects during the hiring process. These advanced studies not only provide deeper insights into the technical and medical aspects of firefighting but also demonstrate a commitment to the profession and a readiness to take on leadership roles in the future.

Furthermore, certifications and courses in areas such as EMT (Emergency Medical Technician) or paramedic training, hazardous materials handling, and fire safety can make a candidate more attractive to fire departments. These qualifications show a proactive approach to acquiring skills that are directly applicable to firefighting and rescue operations, giving individuals an edge in a competitive field. It's important for candidates to research the specific requirements and preferences of the fire departments they wish to join, as some may value certain certifications or educational backgrounds more highly than others.

In addition to formal education and certifications, candidates should consider engaging in volunteer work or internships with fire departments or emergency services. Such experiences can provide invaluable hands-on learning opportunities and exposure to the realities of the profession. They also offer the chance to build relationships with experienced firefighters and other emergency service professionals, who can provide mentorship, advice, and support throughout the preparation and application process.

Ultimately, while a high school diploma or GED is the minimum educational requirement, aspiring firefighters should view additional education and training as investments in their future careers. By expanding their knowledge and skills beyond the basics, candidates can significantly improve their chances of success in the hiring process and prepare themselves for a rewarding career in firefighting.

2. Firefighter Entrance Exam

Given the pivotal role of the Firefighter Entrance Exam in determining a candidate's eligibility for further progression in the firefighting career path, it is essential to delve into the specifics of the exam's components and the strategies for mastering them. The exam not only tests the candidate's intellectual and physical capabilities but also evaluates their psychological readiness for the demands of firefighting. A comprehensive understanding of these components will enable candidates to tailor their preparation effectively.

Reading Comprehension is a segment of the exam, requiring candidates to swiftly interpret and analyze complex texts.

Mathematical Reasoning challenges candidates with problems that simulate real-life firefighting scenarios, such as calculating the force of water needed to extinguish a fire. Regular practice with algebra, geometry, and basic arithmetic, focusing on word problems and practical applications, can significantly improve performance in this area.

Mechanical Aptitude tests understanding of basic mechanical and physical principles. Candidates can benefit from hands-on experience with tools and machinery, as well as studying the principles that underlie their operation.

Situational Judgment questions assess the ability to make quick and effective decisions in emergency situations. Engaging in role-play scenarios, studying past case studies of emergency responses, and participating in simulation exercises can sharpen these critical thinking skills.

Personality Assessment aims to gauge whether a candidate's personal traits align with the professional demands of firefighting. Honesty and self-awareness are crucial; understanding the core values of the firefighting profession and reflecting on personal experiences can help candidates present themselves authentically and positively.

For the **Physical Ability Test**, a regimen that includes cardiovascular training, strength conditioning, and agility exercises is indispensable. Candidates should familiarize themselves with the specific events included in the test and replicate them as closely as possible in their training. This not only builds the necessary physical strength but also reduces the element of surprise on test day.

Joining a **Study Group** or enrolling in a **Preparation Course** can provide structured learning, accountability, and moral support. These groups offer a platform to exchange knowledge, share resources, and simulate exam conditions, which can alleviate anxiety and boost confidence.

Test-Taking Strategies such as time management, understanding the scoring system, and techniques for eliminating incorrect answers can also enhance exam performance. Candidates should practice these strategies under timed conditions to mimic the pressure of the actual exam.

Feedback and Reflection Candidates should seek constructive feedback from peers, mentors, or through professional review services. Reflecting on performance and adjusting study strategies accordingly can lead to continuous improvement and a higher likelihood of success.

Incorporating **Relaxation and Stress Management Techniques** into the preparation process can help maintain mental and emotional equilibrium. Practices such as meditation, deep breathing exercises, and maintaining a healthy work-life balance can mitigate the stress associated with intense exam preparation.

Community Engagement and **Volunteer Work** not only enrich a candidate's application but also provide practical experience and exposure to the realities of firefighting. These activities underscore a candidate's commitment to public service and can offer unique learning opportunities outside the confines of traditional study materials.

3. Background Check

The background check is a crucial step in the firefighter hiring process, designed to verify the integrity and reliability of candidates. This comprehensive review encompasses criminal records, driving history, and previous employment verification. Fire departments aim to hire individuals who demonstrate not only physical and mental aptitude but also a strong moral character and a sense of responsibility. Given the trust and safety responsibilities placed on firefighters, departments meticulously assess each candidate's past behavior as an indicator of future performance.

Criminal History: Departments will conduct a thorough search of local, state, and federal databases to uncover any criminal records. It's important for candidates to be aware that certain offenses may disqualify them from consideration, depending on their nature and the time elapsed since their occurrence. Honesty is paramount; attempting to conceal past offenses is more detrimental than the records themselves. Candidates with a history of serious or violent crimes are unlikely to progress in the hiring process, reflecting the profession's emphasis on trustworthiness and ethical conduct.

Driving Records: A clean driving record is often a prerequisite for employment as a firefighter. Given that driving emergency vehicles is a significant part of the job, departments scrutinize candidates' driving histories for evidence of reckless or irresponsible behavior. Frequent traffic violations, DUI/DWI convictions, or a history of accidents can raise concerns about a candidate's judgment and reliability. Maintaining a good driving record is thus essential for aspiring firefighters, underscoring the importance of responsible behavior both on and off duty.

Employment History: Previous employment checks are conducted to gauge a candidate's work ethic, reliability, and interpersonal skills. Fire departments contact former employers to verify employment dates, roles, and reasons for leaving, while also seeking insights into the candidate's performance and conduct. A history of job hopping, unexplained gaps in employment, or negative references can be red flags, suggesting issues with commitment, teamwork, or professionalism. Conversely, positive references and a stable work history support a candidate's application, indicating a strong work ethic and the ability to collaborate effectively in high-pressure environments.

Candidates should prepare for the background check by ensuring accuracy in their application materials, addressing any potential concerns upfront, and gathering documents or references that may support their candidacy. Engaging in volunteer work, community service, or other activities that demonstrate a commitment to ethical behavior and public service can also be beneficial. Ultimately, the background check serves to ensure that those selected to serve as firefighters are not only capable of performing the duties of the job but also embody the values and integrity expected of the profession.

4. Medical Examination

The medical examination is part of the firefighter selection process, designed to ensure that candidates are physically capable of performing the demanding tasks required in the profession. Firefighting is an extremely strenuous and physically challenging job that often involves heavy lifting, prolonged physical exertion, exposure to extreme heat, and the need for quick, precise movements in dangerous situations. As such, fire departments need to ensure that prospective firefighters are in peak physical health to handle the rigors of the job.

The medical exam is comprehensive and typically includes evaluations of cardiovascular fitness, lung capacity, musculoskeletal strength, and overall physical health. Below are the key components often covered during this examination:

- **Cardiovascular Endurance**: This is typically assessed through tests such as the 1.5-mile run or step tests, measuring how well a candidate's heart and lungs can perform under sustained physical exertion. Since firefighting often requires long periods of intense activity, strong cardiovascular health is essential.

- **Pulmonary Function**: Firefighters are frequently exposed to smoke and other toxic substances, making lung capacity and respiratory health vital. Pulmonary function tests may involve blowing into a spirometer to measure the volume and efficiency of the lungs.
- **Muscular Strength and Endurance**: Firefighting tasks such as carrying heavy equipment, lifting injured individuals, and breaking through obstacles require considerable strength. The exam typically includes tests for upper and lower body strength, often using weight lifting or specific strength exercises like push-ups, pull-ups, or lifting weighted objects.
- **Flexibility and Joint Health**: Firefighters must maintain agility and flexibility to move quickly and efficiently in tight spaces. The medical exam may include flexibility tests like the sit-and-reach test to assess joint mobility and range of motion, ensuring candidates can maneuver in confined or awkward spaces without risking injury.
- **Vision and Hearing Tests**: Clear vision and hearing are essential for identifying hazards, reading instruments, and communicating with team members in high-stress environments. Vision tests typically assess distance vision, peripheral vision, and depth perception, while hearing tests check for the ability to detect a wide range of sounds, including alarms and verbal commands.
- **Overall Physical Health**: A thorough examination of the candidate's general health is performed, including a review of medical history, a physical examination by a physician, and possibly blood and urine tests to screen for underlying conditions that could affect job performance. This also ensures there are no pre-existing conditions that might disqualify candidates, such as severe allergies, chronic respiratory conditions, or musculoskeletal issues.

In addition to these tests, the medical examination serves as an opportunity to educate candidates about maintaining long-term physical fitness. Firefighters are required to maintain a high level of physical health throughout their careers, and many departments will provide guidance on fitness regimes and healthy living practices to help candidates prepare for the ongoing demands of the job.

6. Drug Screening

The drug screening process is an essential step for all aspiring firefighters, designed to ensure that candidates are free from illegal substances. This procedure reflects the high standards of physical fitness, responsibility, and integrity required in the firefighting profession. Given the demanding nature of firefighting duties, which often involve life-saving responsibilities and operating heavy machinery, it is imperative that all candidates maintain a clear state of mind and optimal physical health. The drug screening typically involves a urine test, although some departments may also require blood tests or hair follicle tests for a more comprehensive analysis of a candidate's substance use history.

Candidates should be aware that the screening process can detect a wide range of substances, including but not limited to marijuana, cocaine, opiates, amphetamines, and phencyclidine (PCP). It is important to note that the legalization of certain substances, such as marijuana in some states, does not exempt candidates from being tested or disqualified if detected. Fire departments adhere to federal laws and regulations, which still classify marijuana as an illegal substance. Therefore, candidates are strongly advised to abstain from all illegal drug use well in advance of the application process.

Preparation for the drug screening should begin as early as possible. Candidates are encouraged to lead a healthy lifestyle, which includes regular exercise, a balanced diet, and adequate hydration. These practices not only contribute to passing the drug test but also to the overall physical fitness required for the job. Additionally, candidates should be mindful of over-the-counter medications and supplements that could potentially lead to false positives on drug tests. It is advisable to disclose any medications or supplements being taken when undergoing the drug screening.

Understanding the implications of a failed drug test is crucial. A positive result can lead to immediate disqualification from the hiring process. In some cases, candidates may be allowed to reapply after a certain period, typically ranging from six months to a year, provided they can demonstrate that they are drug-free. However, this is not guaranteed and varies by department. The best approach is to ensure a clean drug test result by abstaining from illegal substances and adhering to a healthy lifestyle.

In summary, the drug screening is a component of the firefighter hiring process, underscoring the profession's commitment to safety, integrity, and public trust. Candidates should approach this step with the seriousness it deserves, recognizing that their eligibility for a career in firefighting depends on a clean result. By maintaining a drug-free lifestyle and preparing adequately for the screening, candidates can move forward confidently in their journey to becoming firefighters.

7. Psychological Evaluation

The psychological evaluation is another step in the firefighter selection process, designed to assess a candidate's mental and emotional stability. This evaluation ensures that candidates are capable of handling the high-stress situations they will encounter on the job. Firefighting is not only physically demanding but also places a significant mental and emotional burden on individuals. The ability to remain calm under pressure, make quick decisions in emergency situations, and cope with the potential trauma of tragic events is essential for every firefighter. During the psychological evaluation, candidates may undergo a series of tests and interviews with psychologists or psychiatrists specializing in occupational health. These assessments aim to identify any underlying psychological conditions that could impair a candidate's ability to perform firefighting duties effectively. It's important for candidates to approach these evaluations with honesty and openness. The objective is not to exclude candidates with any history of mental health issues but to ensure that they possess the resilience, coping mechanisms, and support systems necessary to thrive in this challenging profession. Additionally, the evaluation seeks to confirm that candidates have the emotional intelligence to work effectively in team settings, communicate under stress, and show empathy towards victims and their families. Preparing for the psychological evaluation involves self-reflection and possibly seeking feedback from peers or mentors on one's emotional and mental strengths and weaknesses. Engaging in stress-reduction techniques, such as mindfulness, exercise, or talking therapies, can also be beneficial. Understanding one's own reactions to stress and developing healthy coping strategies are valuable skills not only for passing the psychological evaluation but for a successful career in firefighting. Remember, mental and emotional health is as important as physical fitness in the firefighting profession. Candidates are encouraged to view the psychological evaluation not as a hurdle but as an opportunity to gain insights into their suitability for a career that demands both physical and psychological resilience.

8. Oral Interview Evaluation

The oral interview is designed to assess a candidate's communication skills, decision-making abilities, and overall suitability for the demanding role of a firefighter. This stage goes beyond evaluating technical knowledge and physical capabilities; it delves into the candidate's personality, ethics, and ability to work under pressure. **Preparation** for the oral interview is key. Candidates should familiarize themselves with common interview questions and practice their responses. It's not just about providing the right answers but also demonstrating how you communicate those answers effectively. **Active listening** is crucial; pay attention to the questions asked and make sure your responses are well-structured and concise. **Scenarios** are often used during the interview to evaluate decision-making skills. You might be asked how you would handle specific emergency situations. This is where your ability to think on your feet and apply your knowledge and training to real-world scenarios is tested. **Teamwork** questions assess your ability to work collaboratively within a team. Firefighting is a team effort, and your responses should reflect an understanding of the importance of communication, mutual respect, and support among team members. **Ethical dilemmas** may also be presented to gauge your moral compass and integrity. How you handle these questions can tell the panel a lot about your character and values. **Feedback** from mock interviews can be invaluable. Seek out mentors or peers who can provide constructive criticism on your interview technique, body language, and even your attire. Remember, the goal is to present yourself as a competent, confident candidate who is ready for the challenges of firefighting. **Research** the specific fire department you're applying to. Understanding their values, community involvement, and the services they provide can help you tailor your responses and demonstrate a genuine interest in becoming a part of their team. **Non-verbal communication** also plays a significant role. Your posture, eye contact, and the way you manage your nervousness can leave a lasting impression. Finally, **questions for the panel** are usually welcomed at the end of the interview. This is an opportunity to show your enthusiasm for the role and the department. Prepare thoughtful questions that reflect your interest in growth, training opportunities, and community engagement. The oral interview is not just about assessing your suitability for the role but also about you evaluating if the department is the right fit for you.

9. Polygraph Test

The polygraph test, commonly referred to as a lie detector test, is a tool some fire departments use during the hiring process to verify the accuracy of a candidate's background information. This step is crucial as it helps ensure the integrity and reliability of potential firefighters, who must often work in situations demanding high levels of trust and ethical standards. During the polygraph examination, candidates are asked a series of questions related to their application and background, including any past criminal activities, drug use, and other behaviors that could impact their suitability for the role of a firefighter. The test measures physiological responses such as heart rate, blood pressure, respiratory rate, and skin conductivity while the candidate answers these questions. It's important for candidates to approach the polygraph test with honesty and openness. Being nervous is natural, but lying can increase physiological responses, which may be interpreted as deceptive behavior. Preparation for the polygraph test involves understanding the types of questions that may be asked and reflecting on one's past behaviors and experiences. Candidates should ensure they are well-rested and calm before taking the test, avoiding caffeine or other substances that could affect their physiological responses. Remember, the goal of the polygraph test is not to disqualify candidates but to confirm the truthfulness of the information they have provided. Fire departments value integrity and honesty highly, traits that are essential for the safety and well-being of both firefighters and the communities they serve.

10. Probationary Period

The probationary period is another phase in the journey to becoming a firefighter, serving as both an extension of the selection process and an integral part of on-the-job training. During this time, candidates are closely monitored to assess their adaptability to the demands of the profession, their ability to apply their training in real-world scenarios, and their overall fit within the fire department's culture and team dynamics. This period typically lasts between six months to one year, depending on the department's policies and the candidate's performance.

Performance Evaluation is a continuous process throughout the probationary period, focusing on practical skills, knowledge application, and adherence to departmental protocols. Probationary firefighters are expected to demonstrate competence in a wide range of tasks, from firefighting techniques and emergency medical services to routine station duties and community engagement activities. Regular feedback sessions with supervisors and mentors provide opportunities for growth and improvement, highlighting areas of strength and addressing any weaknesses.

Training and Education continue during the probationary period, with probationary firefighters participating in advanced training exercises, specialized courses, and continuous learning opportunities. This ongoing education is crucial for developing the comprehensive skill set required for successful firefighting and emergency response. It also prepares the probationary firefighter for future career advancement opportunities within the department.

Mentorship and Team Integration play a significant role in the probationary firefighter's development. Experienced firefighters often serve as mentors, offering guidance, support, and insight into the nuances of the profession. This relationship facilitates the probationary firefighter's integration into the team, promoting a sense of belonging and enhancing teamwork skills. Effective communication, mutual respect, and collaboration are emphasized, reinforcing the department's commitment to a cohesive and supportive work environment.

Evaluation of Conduct and Professionalism is paramount, with probationary firefighters expected to uphold the highest standards of integrity, ethical behavior, and professionalism. This includes punctuality, respect for colleagues and community members, and a proactive approach to responsibilities. The probationary period is a time for candidates to prove their commitment to the values and mission of the fire service, demonstrating their readiness to serve as trusted and reliable members of the firefighting team.

Throughout the probationary period, candidates are evaluated on their ability to handle the physical, emotional, and intellectual challenges of the firefighting profession. Successful completion of this phase is a testament to a candidate's resilience, skill, and dedication, marking their transition from probationary status to full-fledged firefighters. It's a time of significant personal and professional growth, laying the foundation for a rewarding career in the fire service.

11. Fire Academy Training

Fire Academy Training is a pivotal phase in the journey to becoming a firefighter, blending rigorous classroom instruction with hands-on practical training. This comprehensive approach ensures that candidates not only grasp the theoretical underpinnings of fire science but also acquire the skills necessary for effective firefighting and emergency response. The curriculum is meticulously designed to cover a wide array of topics, from the basics of fire behavior and the use of personal protective equipment (PPE) to advanced rescue techniques and emergency medical services (EMS). Each module is structured to build upon the previous, facilitating a progressive learning experience that prepares candidates for the multifaceted challenges of the firefighting profession.

Classroom instruction provides the foundational knowledge essential for understanding fire dynamics, fire prevention methods, and the safety protocols that safeguard firefighters and the communities they serve. Instructors, who are often experienced firefighters themselves, deliver lectures on fire chemistry, building construction, fire code regulations, and the principles of fire suppression. This theoretical grounding is crucial, as it informs the decision-making process during emergency operations, enabling firefighters to quickly assess situations and deploy appropriate tactics.

The practical training component immerses candidates in real-world scenarios, simulating the high-pressure environment of fire and rescue operations. This hands-on approach is conducted in controlled settings that mimic residential, commercial, and industrial fire situations. Candidates learn to operate firefighting equipment, such as hoses, ladders, and breathing apparatus, under the watchful eye of their instructors. They engage in live fire drills, search and rescue operations, and vehicle extrication exercises, gaining the proficiency and confidence needed to perform under the stress of actual emergencies. The emphasis on teamwork and communication is paramount throughout this training, as effective coordination and collaboration are key to successful firefighting operations.

Safety is the cornerstone of fire academy training, with rigorous protocols in place to protect candidates from injury. However, the inherently physical nature of firefighting is reflected in the training regimen, which includes physical fitness assessments and exercises designed to build strength, endurance, and agility. This physical preparation is complemented by instruction in self-care and stress management techniques, acknowledging the mental and emotional demands of the profession.

As candidates progress through the fire academy, they are continually evaluated on their knowledge, skills, and abilities. These assessments are meant to ensure that only those who meet the high standards of the profession advance to the next stage of their firefighting career. The journey through fire academy training is challenging, yet it is also a time of significant personal and professional growth. Candidates develop not only the technical skills required for firefighting but also the resilience, teamwork, and leadership qualities that define the best in the profession.

Upon completing the rigorous fire academy training, candidates emerge with a deep understanding of fire behavior, emergency medical procedures, hazardous material handling, and the incident command system that orchestrates the response to fires and other emergencies. This comprehensive education ensures that graduates are not only technically proficient but also adaptable to the evolving nature of fire and emergency services. The curriculum is designed to instill a lifelong learning ethos, encouraging firefighters to pursue further education and specialized training throughout their careers. This commitment to continuous improvement is essential for staying abreast of the latest fire science research, technological advancements in firefighting equipment, and evolving emergency response tactics.

The integration of real-life case studies and guest lectures from seasoned fire service professionals enriches the academy experience, providing candidates with insights into the challenges and rewards of the firefighting profession. These sessions offer valuable lessons on leadership, ethical decision-making, and the importance of community engagement in fire prevention and safety education. By understanding the broader context in which firefighting operates, candidates are better prepared to serve as ambassadors of fire safety in their communities, advocating for practices that reduce fire risk and enhance public safety.

Collaboration with local fire departments and emergency medical services during training fosters a sense of camaraderie and professional networking that is invaluable for career development. These partnerships often provide opportunities for internships and mentorship, allowing candidates to gain experience and guidance from working professionals. This real-world exposure complements the academy's curriculum, bridging the gap between theoretical knowledge and practical application.

Upon graduation from the fire academy, candidates are equipped with the certifications and qualifications required to begin their careers as probationary firefighters. This milestone marks the beginning of their journey in the fire service, a career characterized by dedication, bravery, and a commitment to protecting life and property. The transition from academy training to active duty is supported by ongoing professional development opportunities, including advanced training courses, leadership development programs, and specialized certifications that enable firefighters to enhance their skills and advance in their careers.

The fire academy experience is transformative, molding individuals into professionals who are prepared to face the challenges of modern firefighting with expertise, integrity, and compassion. Graduates leave the academy not only as skilled firefighters but as integral members of the emergency response community, ready to make a meaningful impact in their chosen profession.

Chapter 2: Overview of Firefighter Exams

The Firefighter Entrance Exam is a pivotal step in the journey to becoming a firefighter in the United States. This exam is designed to assess a candidate's readiness and suitability for a career in firefighting, covering a broad range of skills and knowledge areas. It's important for candidates to understand that the structure and content of these exams can vary significantly from one location to another, reflecting the specific needs and priorities of different fire departments. At its core, the exam typically includes sections on reading comprehension, mathematical reasoning, mechanical reasoning, situational judgment, and personality assessment. Each of these sections aims to evaluate different facets of a candidate's capabilities, from their ability to process and understand written information to their aptitude for solving practical problems and making sound decisions in stressful situations.

Reading Comprehension tests a candidate's ability to quickly read, understand, and interpret written material. This section is crucial, as firefighters must often follow written instructions and understand complex documents such as fire codes and safety protocols. **Mathematical Reasoning** assesses basic mathematical skills, including arithmetic, algebra, and geometry, which are essential for tasks such as calculating water flow rates or determining the correct angles for ladder placement. **Mechanical Reasoning** examines a candidate's understanding of basic mechanical and physical principles, reflecting the practical, hands-on nature of firefighting work. This might include questions on levers, pulleys, or gears, which are relevant to the operation of firefighting equipment.

Situational Judgment tests present hypothetical scenarios that a firefighter might encounter on the job, asking candidates to choose the most appropriate response from a set of options. These questions assess judgment, problem-solving skills, and the ability to prioritize actions under pressure. **Personality Assessment** sections are designed to gauge a candidate's suitability for the teamwork-oriented and high-stress environment of firefighting. They may explore traits such as integrity, dependability, and cooperation.

Preparation for the Firefighter Entrance Exam requires a comprehensive approach. Candidates are advised to familiarize themselves with the exam format of their target department, as this can influence their study strategy. Many fire departments offer study guides or sample questions, which can provide valuable insights into the specific content and style of questions to expect. Additionally, engaging in regular physical fitness activities is crucial, as some departments incorporate a physical ability test as part of the entrance examination process. This physical component may include tasks like climbing stairs, lifting equipment, or dragging hoses, designed to simulate the physical demands of firefighting work.

It's also beneficial for candidates to engage in study groups or find a mentor who has successfully navigated the process. These resources can offer personalized advice, share study materials, and provide moral support throughout the preparation period. Online forums and social media groups dedicated to firefighting candidates can also be a rich source of information and encouragement.

As the exam date approaches, candidates should focus on areas where they feel less confident, while also maintaining a balanced study plan that covers all sections of the exam. Practice tests can be particularly useful for identifying strengths and weaknesses, allowing candidates to adjust their study focus accordingly. Time management is another skill to develop, as candidates must be able to complete each section of the exam within the allotted time. Practicing under timed conditions can help improve speed and efficiency, reducing the stress of the actual exam day.

Understanding the Firefighter Entrance Exam is just the first step in a comprehensive preparation strategy. With the right resources and a dedicated approach to study and physical fitness, candidates can enhance their chances of success and take a significant step forward in their journey to becoming firefighters.

To optimize preparation efforts, candidates should also consider the psychological aspects of test-taking, such as dealing with anxiety and enhancing concentration.

Techniques like deep breathing, visualization, and positive self-talk can be effective in managing exam stress. Additionally, establishing a consistent study schedule that includes breaks and leisure activities can prevent burnout and maintain mental sharpness.

Community engagement and practical experience can further enrich a candidate's readiness. Volunteering with local fire departments or community emergency response teams offers hands-on experience that can provide a deeper understanding of the firefighting profession. Such involvement not only bolsters a resume but also gives candidates a realistic preview of the challenges and rewards of a firefighting career. It's an opportunity to learn directly from professionals in the field and to apply theoretical knowledge in practical settings.

Moreover, staying informed about the latest developments in fire science and emergency medical services can give candidates an edge. Reading trade journals, attending relevant workshops, and participating in related webinars can expand a candidate's knowledge base and demonstrate a genuine interest in the profession. This proactive approach to learning underscores a commitment to personal and professional growth, qualities highly valued in the firefighting community.

In addition to individual study and physical preparation, seeking feedback from peers and instructors can provide insights into a candidate's performance. Constructive criticism can highlight areas for improvement that may not be apparent to the candidate. It's also an opportunity to refine interpersonal and communication skills, which are crucial for the oral interview portion of the selection process.

Finally, understanding the role of a firefighter in the broader context of community safety and service can inspire candidates to approach their preparation with a sense of purpose and dedication. Reflecting on the impact firefighters have on saving lives and protecting property can be a powerful motivator. It reinforces the importance of thorough preparation, not just for passing the exam, but for embarking on a career that demands excellence, resilience, and a deep commitment to serving others.

The journey to becoming a firefighter is rigorous and demanding, but it is also immensely rewarding. With diligence, perseverance, and a focus on comprehensive preparation, aspiring firefighters can achieve their goals and make a meaningful contribution to their communities.

Chapter 3: Content vs. Skill Focus

Understanding the difference between content-focused and skill-focused questions is pivotal for candidates preparing for the Firefighter Entrance Exam. Content-focused questions are designed to assess a candidate's knowledge of specific information, theories, and concepts relevant to firefighting. These questions might ask about the components of the fire triangle, types of fire extinguishers, or basic first aid procedures. The key to excelling in this area lies in thorough study and memorization of firefighting principles, codes, and safety procedures. Candidates should dedicate time to reading textbooks, attending relevant courses, and reviewing study guides to ensure a comprehensive understanding of the material.

On the other hand, skill-focused questions evaluate a candidate's ability to apply their knowledge in practical, often complex, scenarios. These questions test critical thinking, problem-solving abilities, and the application of knowledge to new situations. For example, a skill-focused question might present a scenario where a candidate must decide the most effective way to use resources at a multi-vehicle accident scene. Preparing for these types of questions requires a different approach. Candidates should practice by working through scenario-based exercises, participating in simulation training, and engaging in group study sessions where they can discuss and debate different strategies for handling emergency situations.

Both types of questions are integral to the Firefighter Entrance Exam, reflecting the dual need for firefighters to have a solid foundation of knowledge and the ability to think on their feet. To prepare effectively, candidates must balance their study time between learning the factual content and developing their analytical and decision-making skills. Engaging with a variety of study materials, including flashcards for memorization, scenario-based practice tests, and interactive simulations, can provide a well-rounded preparation strategy. Additionally, seeking feedback from mentors and experienced firefighters on both content knowledge and practical skills can offer valuable insights and guidance.

It is also beneficial for candidates to familiarize themselves with the specific format and structure of the exam they will be taking. Understanding whether the exam leans more heavily towards content or skill-focused questions, or if it balances both equally, can help in tailoring the preparation strategy accordingly. Joining study groups, forums, or online communities where past and current candidates share their experiences and advice can provide a wealth of information on what to expect and how best to prepare.

In conclusion, excelling in the Firefighter Entrance Exam requires a strategic approach to studying that encompasses both content-focused and skill-focused preparation. By understanding the nature of these questions and adopting a comprehensive study plan, candidates can enhance their chances of success and take an important step towards a rewarding career in firefighting.

Content-Focused Questions

To effectively prepare for content-focused questions on the Firefighter Entrance Exam, candidates must delve deep into the specifics of firefighting knowledge and principles. This entails a rigorous study of fire science, including the chemistry of fire, the behavior of different materials in fire, and the various methods used for firefighting and rescue operations. Understanding the **types of fires**—such as electrical, chemical, and natural—is crucial, as each requires a unique approach for containment and extinguishment. Candidates should be familiar with the **Fire Triangle** (oxygen, heat, and fuel) and how altering any element can control or extinguish a fire.

Knowledge of **firefighting equipment** is another area subject to testing. Candidates must know the names, functions, and maintenance procedures for personal protective equipment (PPE), tools, and apparatus used in firefighting. This includes everything from helmets and boots to hoses, nozzles, and fire extinguishers, as well as the types of fire apparatus like pumpers, ladders, and rescue vehicles.

First aid and CPR are also tested in content-focused questions. Candidates should understand basic life-saving techniques, the proper use of an Automated External

Defibrillator (AED), and how to address common injuries encountered by firefighters and victims at the scene of an emergency.

Fire codes and regulations form a significant part of the exam. Familiarity with national and local fire safety codes, building codes, and the reasons behind these regulations is essential. This knowledge ensures that firefighters can not only respond effectively to emergencies but also work on fire prevention and education within their communities.

To master these areas, candidates should engage in a variety of study methods. Reading textbooks and fire science literature, attending workshops and seminars, and utilizing online resources are all effective strategies. Flashcards can be particularly useful for memorizing terms and definitions, while group study sessions allow for discussion and deeper understanding of complex topics. Practical experience, such as volunteering with a local fire department or participating in fire science programs, can provide invaluable hands-on learning.

Sample Multiple-Choice Questions to illustrate the type of content-focused questions candidates might encounter include:

1. What element is not part of the Fire Triangle?
[A] Oxygen
[B] Carbon
[C] Heat
[D] Fuel

2. Which type of fire extinguisher is most suitable for a grease fire?
[A] Water
[B] Dry Chemical
[C] Carbon Dioxide
[D] Wet Chemical

3. What is the primary purpose of the Incident Command System (ICS)?
[A] To provide a standard for emergency communications

[B] To ensure efficient, effective incident management

[C] To outline the responsibilities of fire investigators

[D] To standardize the equipment used by emergency responders

4. When performing CPR on an adult, at what rate should chest compressions be given?

[A] 60 compressions per minute

[B] 80 compressions per minute

[C] 100 to 120 compressions per minute

[D] 140 compressions per minute

5. Which NFPA standard is known as the Standard on Fire Department Occupational Safety, Health, and Wellness Program?

[A] NFPA 1001

[B] NFPA 1500

[C] NFPA 1710

[D] NFPA 1981

Preparing for content-focused questions requires a comprehensive study plan that covers a wide range of topics. By focusing on the foundational knowledge and understanding the principles behind firefighting practices, candidates can approach the Firefighter Entrance Exam with confidence, ready to demonstrate their mastery of the essential content required for a career in firefighting.

Skill-Focused Questions

To excel in skill-focused questions, candidates must hone their ability to apply theoretical knowledge to practical scenarios, demonstrating proficiency in critical thinking, problem-solving, and decision-making. This segment of the exam assesses not just what you know, but how effectively you can use that knowledge in real-world firefighting situations. It's about showing your capacity to think on your feet, analyze complex situations, and make informed decisions under pressure. To prepare for these types of questions, engaging in hands-on training exercises, participating in simulation drills, and practicing with scenario-based questions are key strategies. These activities help bridge the gap between theoretical knowledge and practical application, enabling candidates to experience firsthand the types of challenges they might face on the job.

Scenario-Based Practice is crucial for mastering skill-focused questions. By immersing yourself in simulated emergency situations, you can develop a deeper understanding of how to apply firefighting principles and techniques in a variety of contexts. Whether it's deciding the best approach to tackle a rapidly spreading fire in a multi-story building or determining the most effective way to conduct a search and rescue operation in a smoke-filled environment, scenario-based practice will enhance your problem-solving skills and improve your ability to make quick, effective decisions.

Critical Thinking Exercises are another essential component of preparing for skill-focused questions. These exercises encourage you to think beyond the immediate problem, consider various outcomes, and evaluate the potential impact of different actions. Developing a habit of asking yourself, "What if?" and considering multiple scenarios and their possible consequences can sharpen your analytical skills and prepare you for the complex decision-making required in the firefighting profession.

Team-Based Discussions and Debates offer valuable opportunities to explore different perspectives and approaches to problem-solving. Engaging with peers in discussions about hypothetical emergency situations allows you to hear different viewpoints, justify your own decisions, and refine your thought process. This collaborative learning experience can broaden your understanding of firefighting tactics and strategies, making you better equipped to tackle skill-focused questions on the exam.

Feedback from Mentors and Experienced Firefighters can provide insights into the practical application of firefighting knowledge. Seeking advice and feedback from those who have firsthand experience in the field can help you identify areas for improvement, understand common pitfalls, and learn effective strategies for dealing with complex situations. Their guidance can be invaluable in helping you develop the practical skills and judgment needed to excel in skill-focused exam questions.

Sample Scenario-Based Multiple-Choice Questions:

1. During a residential fire, you are faced with a choice between attempting to extinguish a small fire in the kitchen or evacuating a family trapped on the second floor. Which action should you prioritize?
[A] Extinguish the kitchen fire first to prevent it from spreading
[B] Evacuate the family trapped on the second floor immediately
[C] Call for backup before taking any action
[D] Attempt to do both simultaneously with available resources

2. You arrive at the scene of a car accident where a vehicle is leaking gasoline. There is no fire present, but there is a risk of ignition. What is the most appropriate initial action?
[A] Use a fire extinguisher on the gasoline leak
[B] Evacuate the area and establish a safety perimeter
[C] Begin extrication of the vehicle's occupants without addressing the leak
[D] Apply water to the leak to dilute the gasoline

3. When responding to a hazardous material spill inside a warehouse, what is the first step you should take upon arrival?
[A] Begin evacuation of the immediate area
[B] Identify the substance involved using available placards or labels
[C] Enter the warehouse to assess the extent of the spill
[D] Contact the hazardous materials team and wait for their arrival

Part B: Fire Science and Behavior

Chapter 4: Fire Science Basics

Understanding the **Fire Triangle** is fundamental to grasping the basics of fire science. This concept illustrates the three essential elements a fire needs to ignite and sustain itself: **oxygen**, **heat**, and **fuel**. Without any one of these components, a fire cannot start or continue to burn. Oxygen, found in the air around us, supports the chemical reactions that occur during combustion. Heat, which can come from various sources such as a match or an electrical fault, provides the energy necessary to start the combustion process. Fuel can be anything combustible, from dry leaves in a forest to the gas in a kitchen stove. By controlling and removing one or more of these elements, firefighters can effectively extinguish fires.

Moving on to **types of fires**, it's crucial to recognize that not all fires are the same and thus, cannot be treated in the same manner. Fires are classified into classes based on the fuel that is burning, which directly affects the method used to extinguish them. **Class A** fires involve ordinary combustibles like wood, paper, and cloth. **Class B** fires are fueled by flammable liquids such as gasoline, oil, or grease. **Class C** fires involve electrical equipment and require non-conductive extinguishing agents to prevent electrical shock. **Class D** fires burn metal and require specific extinguishing agents that do not react violently with the burning material. Lastly, **Class K** fires are a special category for kitchen fires that involve cooking oils and fats. Each class of fire requires different strategies and tools for effective extinguishment, highlighting the importance of firefighters understanding the nature of the fire they are dealing with.

Fire behavior is another area of study, encompassing how fires start, spread, and can be controlled or extinguished. Fire behaves differently depending on a variety of factors, including the type of fuel, the availability of oxygen, and environmental conditions such as wind and humidity. For example, a fire in a closed room will behave differently than a wildfire in an open field due to differences in oxygen availability and fuel types. Understanding fire behavior is essential for predicting how a fire will spread and the best methods for containment and extinguishment.

To effectively fight fires, firefighters must also be familiar with **firefighting equipment**. This includes personal protective equipment (PPE) like helmets, gloves, and turnout gear that protect firefighters from heat, flames, and other hazards. Firefighters also use a variety of tools and apparatus, including hoses, nozzles, ladders, and fire extinguishers, each designed for specific types of fires and situations. Knowledge of how to properly use and maintain this equipment is crucial for both the safety of firefighters and the effectiveness of firefighting efforts.

In addition to equipment, **first aid and CPR** knowledge is a mandatory requirement for firefighters. In many cases, firefighters are the first responders to emergency situations where victims may require immediate medical attention. Being able to perform basic life support measures can be the difference between life and death in these moments.

Fire codes and regulations play a significant role in fire prevention and safety. These rules and guidelines are designed to minimize the risk of fires and ensure that buildings and other structures are constructed and maintained in a way that promotes safety and facilitates effective emergency response. Firefighters must be familiar with these codes to identify potential hazards, enforce regulations, and educate the public on fire safety practices.

To prepare for the variety of content-focused questions on the Firefighter Entrance Exam, candidates should engage in comprehensive study and hands-on practice. This includes reviewing fire science principles, familiarizing themselves with the different types of fires and appropriate extinguishing methods, and understanding the importance of fire codes and regulations. Additionally, practical experience with firefighting equipment and first aid procedures will be invaluable. Through a combination of theoretical knowledge and practical skills, candidates can build a solid foundation for a successful career in firefighting.

Understanding the **Fire Triangle** and the **types of fires** is just the beginning; mastering **fire behavior** and the use of **firefighting equipment** is equally crucial. The way a fire spreads can dramatically change the approach firefighters must take. For instance, the **flashover** phenomenon, a stage of fire development, occurs when nearly all combustible materials in a room ignite almost simultaneously, drastically increasing the fire's intensity and danger.

Another aspect of fire behavior is the **backdraft** scenario, a sudden and explosive increase in fire's intensity caused by the introduction of oxygen into an oxygen-depleted environment. This can happen when a door or window is opened in a burning building. Understanding these behaviors helps firefighters anticipate and mitigate potentially lethal explosions.

The stages of a fire—**incipient**, **growth**, **fully developed**, and **decay**—also provide a framework for understanding how fires evolve over time. The **incipient stage** represents the fire's ignition point, where it might still be controlled or extinguished with minimal equipment. During the **growth stage**, the fire begins to spread, and temperatures rise significantly, making direct attack more challenging. The **fully developed stage** is characterized by the fire consuming all available fuel, reaching peak temperatures. Finally, the **decay stage** occurs as the fire runs out of fuel or is extinguished, significantly reducing in size and intensity.

For effective firefighting, knowledge of **firefighting equipment** and **first aid and CPR** is indispensable. **Personal Protective Equipment (PPE)**, such as self-contained breathing apparatus (SCBA), protects firefighters from smoke inhalation and heat, while **thermal imaging cameras** can help locate victims and hotspots through smoke and darkness. Familiarity with the operation and maintenance of **fire apparatus**, including engines, ladders, and specialized tools, ensures that firefighters can efficiently respond to and manage fire scenes.

First aid and CPR skills are essential, as firefighters often provide life-saving care before medical professionals arrive. Skills such as performing CPR, treating burns, and managing trauma injuries can save lives in the moments following an emergency.

Lastly, an understanding of **fire codes and regulations** is essential not only for preventing fires but also for ensuring that firefighting operations comply with legal and safety standards. These regulations cover a wide range of topics, from building design and construction materials to fire alarm systems and emergency exit requirements.

Firefighters play a key role in enforcing these codes, conducting inspections, and educating the public on fire safety measures.

In preparing for the Firefighter Entrance Exam, candidates must immerse themselves in both the theoretical and practical aspects of firefighting. This includes not only memorizing facts and procedures but also engaging in hands-on training to develop the physical skills and instincts necessary for effective firefighting. By combining a deep understanding of fire science with practical experience and a commitment to ongoing learning, candidates can position themselves for success on the exam and in their future careers as firefighters.

Chapter 5: Advanced Fire Behavior

The phenomena of flashover and backdraft, along with the stages of fire development, represent some of the concepts that candidates must grasp.

Flashover is a pivotal event in the development of a compartment fire, marking the transition from a localized fire to a full-room involvement. This occurs when the room's contents and gases reach ignition temperature almost simultaneously, resulting in a rapid and widespread fire that can engulf an entire room or area. Understanding the conditions that lead to a flashover is essential for firefighters to anticipate and prevent these dangerous scenarios. Recognizing signs such as rapidly increasing heat, darkening smoke, and rollover (the appearance of flames across the ceiling) can help firefighters predict and avoid being caught in a flashover.

Backdraft presents a different but equally hazardous situation. It happens when a fire has consumed most of the available oxygen within a space and is suddenly reintroduced to fresh oxygen, typically when a door, window, or other opening is made. The result is a violent explosion of fire, posing significant risk to firefighters and anyone nearby. Identifying potential backdraft conditions, such as windows that appear to be breathing or smoke puffing from cracks, is necessary for preventing injuries or fatalities.

The **stages of fire** offer a framework for understanding how a fire behaves from its inception through to its decline. The **incipient stage** is the initial phase, where the fire begins with a heat source igniting available fuel. This stage offers the best chance for extinguishment with minimal damage. As the fire progresses to the **growth stage**, it starts to spread to nearby combustibles, and the heat **release rate increases**.

In firefighting and fire science, the term **"release rate"** often refers to the **Heat Release Rate (HRR)**, which is the rate at which heat energy is produced by a fire. When we say "release rate increases," we mean that the amount of heat generated per unit of time by the burning materials is growing.

This increase in the release rate can occur for several reasons. As a fire develops, more combustible material, or fuel, may ignite, allowing the fire to produce more heat. For example, a small flame can quickly spread to nearby flammable objects, resulting in a higher overall energy output and, consequently, a higher release rate. Additionally, the availability of oxygen plays a significant role. Fires require oxygen to sustain combustion, and when fresh air flows into an enclosed space—perhaps when a door or window is opened—the fire gains more oxygen. This boost in oxygen can increase combustion efficiency and contribute to a higher heat release rate.

The release rate is also influenced by the fire's growth stage. Fires typically progress through four main stages: ignition, growth, fully developed, and decay. During the growth stage, the heat release rate rises significantly as the fire spreads and consumes more fuel. This increasing release rate indicates that the fire is becoming more intense, which can make it harder to control.

An increase in the release rate has serious implications. For instance, a higher release rate can lead to **flashover**—a sudden and dangerous transition where nearly all combustible surfaces in a room ignite simultaneously due to the intense heat buildup. This stage is especially hazardous, as it can rapidly intensify the fire's spread and make firefighting efforts more challenging. Additionally, the increased thermal load on a building's structure due to the higher release rate can weaken materials and potentially lead to structural collapse.

For firefighters, a higher release rate signifies more extreme conditions. The intense heat, rapid fire spread, and dense smoke can severely reduce visibility, complicating both the rescue operations and fire suppression efforts. Understanding and monitoring the release rate of a fire is thus critical for determining the appropriate firefighting strategies to protect property and save lives.

Firefighters arriving during this stage are faced with a rapidly escalating situation requiring immediate action to prevent further spread.

Transitioning into the **fully developed stage**, the fire has reached its peak, consuming all available fuel and oxygen within the area. Temperatures are at their highest, making direct attack challenging and dangerous. Finally, the **decay stage** occurs as the fire runs out of fuel or oxygen, leading to a decrease in intensity. However, hidden hotspots and the potential for re-ignition remain concerns that firefighters must address.

Each of these stages and phenomena requires a specific approach and understanding to manage effectively. For instance, combating a fire in the growth stage may involve different tactics and equipment than those used during the fully developed stage. Similarly, recognizing the signs of flashover or backdraft enables firefighters to take preventive measures, such as ventilating in a controlled manner or applying water from a safe distance.

In addition to theoretical knowledge, practical training and simulations play a crucial role in preparing firefighter candidates to handle these advanced fire behaviors. Engaging in live fire training exercises, utilizing fire behavior simulators, and studying case studies of past fire incidents are all valuable methods for gaining a deeper understanding of these complex phenomena. Through a combination of classroom learning and hands-on experience, candidates can develop the skills and judgment necessary to safely and effectively respond to fires, enhancing their preparedness for both the entrance exam and their future careers in firefighting.

To effectively manage and mitigate the dangers associated with advanced fire behavior, firefighters must be adept at applying water in a manner that cools the environment without causing a steam explosion, which could exacerbate the situation, particularly during a flashover. This requires a precise understanding of nozzle techniques and the physics of water conversion to steam. For backdraft scenarios, the ability to perform controlled ventilation can be lifesaving, reducing the risk of explosion by gradually introducing oxygen to the fire environment in a managed way.

Moreover, the knowledge of fire dynamics is crucial for predicting the movement and growth of a fire within a structure. This includes understanding how building construction and design, ventilation paths, and the presence of combustible materials can influence fire behavior. Firefighters trained in these areas can more effectively strategize their approach to firefighting operations, choosing tactics that limit fire growth and spread, such as isolating the fire or creating firebreaks.

The role of modern technology in understanding and combating advanced fire behavior cannot be understated. Thermal imaging cameras, for instance, allow firefighters to identify hotspots and areas of intense heat through smoke and darkness, guiding them in directing their efforts where they are most needed. Similarly, drone technology can provide aerial views of the fire scene, offering insights into the fire's extent and behavior that would not be possible from the ground.

Fire investigation also plays a part in understanding fire behavior. Post-incident analysis helps in identifying the cause of the fire, the conditions that contributed to its spread, and the effectiveness of the firefighting tactics used. This information is invaluable for improving fire safety measures, developing better firefighting techniques, and preventing future incidents.

For those preparing for the Firefighter Entrance Exam, mastering the concepts of advanced fire behavior is essential. Study materials should include textbooks on fire science, technical manuals on firefighting tactics, and resources on building construction and fire dynamics. Additionally, candidates should seek out opportunities for practical experience, such as participating in fire department ride-alongs, attending fire academy training sessions, and engaging in simulated fire scenarios.

In preparing for the exam and a career in firefighting, understanding the theoretical aspects of fire behavior must be complemented with practical skills and experience. This holistic approach ensures that firefighter candidates are not only able to pass their entrance exams but are also prepared to face the challenges of real-world firefighting with competence and confidence. Engaging in continuous learning and training, staying updated on the latest fire science research, and practicing situational awareness on the fireground are all components of a firefighter's ongoing professional development. Through diligent study and practical application, candidates can equip themselves with the knowledge and skills needed to effectively respond to and manage the complex and dynamic nature of fire, safeguarding lives and property.

Part C: Firefighting Equipment

Chapter 6: Personal Protective Equipment

Personal Protective Equipment (PPE) is a component in the arsenal of firefighting gear, designed to safeguard firefighters against the myriad of hazards they face in the line of duty. Understanding the various types of PPE, their specific uses, and proper maintenance practices is essential for candidates aspiring to join the firefighting ranks. The primary goal of PPE is to provide a barrier between the firefighter and the dangers of the fireground, including extreme temperatures, toxic smoke, and structural hazards.

The foundation of firefighter PPE includes turnout gear, also known as bunker gear, which comprises a coat, pants, boots, and gloves. This ensemble is constructed from fire-resistant materials that protect against heat and flame while allowing for mobility and flexibility during operations. The outer shell is designed to repel water and resist abrasions, while the thermal liner and moisture barrier within provide insulation and protection from steam burns.

Helmets are offering protection against falling debris, burns, and impacts. Modern firefighting helmets are equipped with face shields and goggles to protect the eyes from smoke, heat, and flying particles. Additionally, ear flaps and neck protectors are integrated into the helmet design to safeguard against burns and heat exposure to these sensitive areas.

The SCBA provides clean air to the firefighter through a mask, allowing them to breathe in smoke-filled environments. This equipment is composed of a high-pressure tank, regulator, and mask assembly, which must be meticulously maintained to ensure reliability and safety during use.

Firefighting boots are specially designed to offer protection from sharp objects, chemicals, and electrical hazards. Made from durable, fire-resistant materials, these boots feature reinforced toes and soles to shield the feet from punctures and slips. Additionally, the high-top design and insulation protect against water, heat, and cold, ensuring the firefighter's mobility and comfort in various environments.

Gloves are an essential component of PPE, designed to protect the hands from cuts, burns, and chemical exposures while maintaining dexterity and grip. Firefighting gloves are made from fire-resistant materials with reinforced palms and fingers, providing a balance between protection and functionality.

Maintenance of PPE is crucial to ensure its effectiveness and longevity. Regular inspections, cleaning, and repairs are necessary to keep the gear in optimal condition. Firefighters must be trained in the proper care of their PPE, including how to inspect for damage, clean contaminants, and store the equipment correctly to prevent degradation.

In addition to the standard PPE ensemble, firefighters may also utilize specialized equipment for specific scenarios. This includes wildland firefighting gear, which is lighter and designed for mobility over rough terrain, and hazmat suits for incidents involving hazardous materials. Each type of gear is tailored to the unique challenges of the firefighting environment it is intended for, emphasizing the importance of understanding the correct usage and maintenance practices for each piece of equipment.

Understanding and adhering to the recommended maintenance schedules for each piece of PPE is not just about compliance; it's about ensuring personal safety and maximizing the protective capabilities of the gear. For instance, turnout gear should be inspected after each use for signs of damage or wear. Any compromised gear must be repaired or replaced immediately to maintain its protective properties. Similarly, SCBA units require regular checks to ensure the air supply is uncontaminated and the masks maintain a proper seal.

Firefighters must be adept at quickly putting on and removing their gear, even under the stress of an emergency. This proficiency not only impacts the firefighter's safety but also the efficiency of the response team as a whole. Drills and practice sessions should be a regular part of a firefighter's training regimen to reinforce these skills.

Moreover, advancements in PPE technology and design are ongoing, and staying informed about these developments is crucial. New materials and innovations can offer better protection, greater comfort, or enhanced functionality. Fire departments should encourage and facilitate ongoing education and training for their personnel to ensure they are equipped with the most current and effective gear.

The role of PPE in protecting firefighters cannot be overstated. It is a line of defense against the myriad of dangers present in firefighting operations. However, its effectiveness is heavily dependent on proper use, regular maintenance, and a thorough understanding of its capabilities and limitations. Firefighters must be vigilant in the care of their gear, as it plays a significant role in their safety and effectiveness in the line of duty.

In summary, the comprehensive approach to PPE in firefighting encompasses selection, maintenance, training, and continuous learning. Each element is crucial in ensuring that firefighters are well-protected, enabling them to perform their duties safely and effectively. Aspiring firefighters must grasp the importance of PPE, not only for their safety but also as a fundamental aspect of their professional responsibilities. Through diligent care, regular training, and an understanding of the latest advancements in firefighting technology, firefighters can significantly mitigate the risks associated with their challenging and noble profession.

Chapter 7: Tools and Equipment Overview

In the realm of firefighting, the tools and equipment a firefighter has at their disposal are not just accessories but lifelines. These tools are designed to tackle a wide range of emergencies, from extinguishing fires to rescuing victims trapped in debris. The selection of firefighting tools and equipment is vast, each with a specific purpose and application in various firefighting scenarios. Understanding these tools' functions and importance is crucial for every aspiring firefighter, as it prepares them for the challenges they will face in the field.

One of the most fundamental pieces of equipment is the fire hose, a crucial tool for delivering water and other extinguishing agents to the fire. Fire hoses come in different sizes and lengths, with each type designed for specific situations. For example, a larger diameter hose might be used to combat a significant structural fire, providing a high volume of water, while a smaller hose might be more suitable for tight spaces or less intense fires. The nozzle attached to the hose also plays a role, as it can be adjusted to change the water's pressure and spray pattern, allowing firefighters to adapt quickly to the fire's demands.

Another essential tool in the firefighter's arsenal is the axe, which serves multiple purposes on the fireground. Whether breaking through doors or windows for entry or ventilation or cutting through obstacles during search and rescue operations, the axe is a symbol of the firefighter's versatility and strength. Similarly, the Halligan bar, a multipurpose prying tool, is indispensable for forcible entry, demolition, and even as a lever in vehicle extrication scenarios.

Ladders are another cornerstone of firefighting operations, enabling access to elevated or hard-to-reach areas. From the traditional wooden ladders to the more modern, lightweight aluminum or fiberglass models, each type of ladder has its specific use, whether it's for reaching a window on the second story of a building or providing a stable platform for water delivery from above.

Firefighters also rely heavily on thermal imaging cameras (TICs) to see through smoke and darkness, identifying hotspots, hidden fires, or locating victims in smoke-filled environments. This technology has revolutionized firefighting tactics, allowing for more efficient and safer operations.

However, beyond the standard PPE, firefighters must be familiar with specialized equipment such as the Jaws of Life, used in vehicle extrication to cut through metal and free trapped individuals. This hydraulic rescue tool exemplifies the technological advancements that have enhanced the firefighter's ability to save lives.

The importance of understanding and properly maintaining these tools cannot be overstated. Regular training and drills ensure that firefighters can deploy these tools effectively and safely, minimizing risks to themselves and those they are sworn to protect. Each piece of equipment, from the simplest hand tool to the most advanced technological device, plays a role in the complex and dynamic environment of firefighting. As we delve deeper into the specifics of these tools and their applications, it becomes clear that the knowledge and skills required to utilize them are foundational to the success and safety of firefighting operations.

The ventilation fan is another tool, often used to clear smoke from a structure, making it safer for firefighters to operate and easier to locate victims. By expelling hot gases and smoke, ventilation improves visibility and reduces the risk of flashover, a sudden and intense fire caused by the ignition of combustible gases in an oxygen-depleted environment. Proper use of ventilation techniques, combined with the strategic placement of fans, can significantly enhance the effectiveness of firefighting efforts and increase the safety of both firefighters and victims.

In addition to these more visible tools, firefighters also rely on a variety of smaller, yet equally important equipment.

For example, gas detectors help assess the presence of toxic gases or oxygen deficiency in the air, allowing firefighters to recognize potential hazards before entering a dangerous environment. Likewise, ropes and pulleys are useful in rescue operations, especially in high-angle or confined space rescues, where they enable safe lowering or extraction of victims from hazardous situations.

The pike pole, a long pole with a metal spike and hook on one end, is another versatile tool used for a multitude of tasks, including pulling down ceilings to check for fire extension, opening walls for ventilation, or simply as a means to reach and move objects from a distance. Its simplicity belies its utility, making it a staple in the firefighter's toolkit.

Water extinguishers, foam extinguishers, and dry chemical extinguishers each play a role in combating different types of fires, from electrical to flammable liquids. Understanding the appropriate use of each extinguisher type is crucial, as the wrong choice can exacerbate the situation. Training on the various extinguishers available, as well as their application techniques, is a component of a firefighter's education.

The importance of communication devices cannot be overstated in the context of firefighting operations. Handheld radios ensure constant communication between team members, command units, and other emergency services. This coordination is essential for the effective management of resources and personnel during an emergency, allowing for real-time updates and strategic decision-making.

Lastly, the mobile data terminal (MDT) in fire apparatus provides access to information en route to an incident, including maps, building plans, and hydrant locations. This technology enhances situational awareness and allows for more informed decision-making upon arrival at the scene.

Each of these tools and equipment pieces, from the basic to the technologically advanced, forms a part of the intricate puzzle that is firefighting. Mastery over their use, maintenance, and application is essential for the modern firefighter, underscoring the importance of comprehensive training programs that cover not only the physical aspects of firefighting but also the cognitive skills required to select and utilize the appropriate tool for each unique scenario. The synergy between a well-trained firefighter and their tools is what ultimately makes the difference in saving lives and property, highlighting the profound responsibility that comes with the profession and the role of ongoing education and practice in fulfilling this duty effectively.

Chapter 8: Fire Apparatus Types and Roles

Fire apparatus, often referred to as fire engines or fire trucks, are the backbone of the firefighting profession, providing the necessary tools, equipment, and water supply to combat fires and conduct rescue operations. Each type of fire apparatus is designed with specific functions in mind, tailored to meet the diverse needs of fire services across different environments and scenarios. The primary categories include pumpers, aerials, tankers, and rescue units, each playing a crucial role in firefighting and rescue missions.

Pumpers, also known as engines, are perhaps the most recognizable form of fire apparatus. They are equipped with a large pump, water tank, and hoses of various sizes and lengths. The primary function of a pumper is to supply water to the hoses used by firefighters to extinguish fires. These vehicles are also outfitted with a variety of tools and equipment for forcible entry, ventilation, and search and rescue operations, making them versatile units on the fireground.

Aerial units, including ladder trucks and platforms, are essential for reaching high elevations, whether for rescuing individuals from upper stories of buildings or providing a means to deliver water from above to combat rooftop or high-level fires. Aerial apparatus are equipped with large, extendable ladders or platforms, which can be raised or extended to access buildings and other structures. These units often carry specialized tools for ventilation and forcible entry, as well as equipment for high-angle rescues.

Tankers, also known as tenders in some regions, are needed in areas where the water supply is limited or non-existent, such as rural or undeveloped areas. These vehicles carry large quantities of water, often several thousand gallons, to supply pumpers at the scene of a fire. Tankers are especially important for firefighting operations outside the reach of a municipal water system, where access to fire hydrants is not available.

Rescue units are specialized apparatus designed for a wide range of rescue operations, including vehicle extrications, technical rescues, and hazardous materials incidents. These units are stocked with an array of specialized tools and equipment, such as the Jaws of Life, ropes and pulleys for high-angle rescues, and protective gear for hazardous materials handling. Rescue units are often among the first to arrive at the scene of an emergency, providing services to save lives and mitigate the effects of disasters.

In addition to these primary categories, there are also specialized fire apparatus designed for specific purposes, such as brush trucks for wildland firefighting, hazmat units for hazardous materials incidents, and air/light trucks that provide scene lighting and air replenishment for firefighters' self-contained breathing apparatus (SCBA).

The design and capabilities of fire apparatus have evolved significantly over the years, driven by advancements in technology and changes in firefighting tactics. Modern fire engines are equipped with computerized controls, advanced safety features, and efficient pumping systems, enhancing their effectiveness on the fireground. Additionally, the integration of communication and information technology in fire apparatus has improved coordination and response times, enabling firefighters to respond more effectively to emergencies.

Understanding the different types of fire apparatus and their roles in firefighting operations is essential for aspiring firefighters, as it provides insight into the complexities of the profession and the tools available to combat fires and save lives. The selection and deployment of fire apparatus depend on various factors, including the nature of the emergency, the environment, and the resources available, highlighting the importance of strategic thinking and decision-making in the firefighting profession.

The strategic deployment of these various fire apparatus types is an aspect of firefighting operations, requiring thorough planning and coordination. Incident commanders must assess the situation quickly, determining which types of apparatus are needed based on the specific conditions of each incident. For example, a high-rise fire might necessitate the dispatch of multiple aerial units for rescue and water delivery, while a wildfire would require brush trucks and tankers to navigate rough terrain and provide sufficient water supply. This decision-making process is crucial to ensure the safety of firefighters and civilians, as well as to maximize the effectiveness of the response effort.

Training for firefighters includes extensive instruction on the operation and capabilities of each type of fire apparatus. Firefighters must be familiar with the mechanics and functions of the vehicles they operate, from driving pumpers and aerial units to managing the water supply from tankers. This knowledge allows them to adapt to the dynamic nature of fire incidents, where conditions can change rapidly and unexpectedly. Additionally, firefighters learn to perform maintenance and safety checks on their apparatus, ensuring that all equipment is in optimal condition and ready for immediate deployment.

The role of fire apparatus extends beyond firefighting and rescue operations. Many units are equipped with medical supplies and advanced life support equipment, allowing firefighters to provide emergency medical care at the scene of an incident. This dual role of fire apparatus as both a firefighting and medical response unit underscores the multifaceted nature of the firefighting profession, and the comprehensive training required to serve effectively.

Community outreach and education are also facilitated through the use of fire apparatus. Fire departments often conduct demonstrations and safety talks at schools, community centers, and public events, using their vehicles as teaching tools. These engagements provide an opportunity to educate the public about fire safety, the functions of different fire apparatus, and the role of firefighters in the community. Such interactions not only foster a positive relationship between the fire service and the public but also raise awareness about the importance of fire prevention and preparedness.

The evolution of fire apparatus continues as advancements in technology and engineering emerge. Manufacturers are constantly seeking ways to improve the efficiency, safety, and environmental impact of these vehicles. Innovations such as electric-powered fire engines and drones for aerial surveillance and assessment are examples of how the industry is adapting to meet the future needs of firefighting. As these technologies develop, training programs and operational protocols will also evolve, ensuring that firefighters remain equipped with the best tools to protect lives and property.

In essence, the diversity and complexity of fire apparatus reflect the challenging and dynamic environment in which firefighters operate. Each vehicle, with its specialized equipment and capabilities, plays an indispensable role in the overall strategy of fire suppression and rescue. Aspiring firefighters must gain a deep understanding of these tools of the trade, as their proficiency in using them can make the difference between life and death. The commitment to continuous learning and adaptation is a hallmark of the firefighting profession, driven by the goal of safeguarding the community from the ravages of fire and disaster.

Part D: Emergency Medical Knowledge

Chapter 9: Basic First Aid and CPR

First aid and CPR are essential skills for every firefighter, as they often find themselves in situations where providing immediate medical assistance can mean the difference between life and death.

1. The Importance of First Aid and CPR in Firefighting

Firefighters are often the first responders to a wide range of emergencies, from fires and accidents to medical incidents like heart attacks or traumatic injuries. In these situations, having the ability to provide immediate medical care ensures that victims receive help as quickly as possible, improving their chances of survival. First aid and CPR knowledge is essential not only for treating victims but also for stabilizing them until paramedics or other medical professionals arrive on the scene.

2. CPR: Saving Lives in Cardiac Emergencies

Cardiopulmonary resuscitation (CPR) is a life-saving procedure used when a person's heart stops beating or when they are not breathing. CPR involves a combination of chest compressions and rescue breathing, which helps maintain circulation and oxygen flow to the brain and vital organs until advanced medical care can be provided.

The key steps in performing CPR are as follows:

- **Assess the Situation**: Ensure the scene is safe before approaching the victim. Check if the person is responsive and breathing. If they are unresponsive and not breathing, begin CPR immediately.
- **Chest Compressions**: Place your hands in the center of the victim's chest and deliver firm, quick compressions at a rate of about 100-120 compressions per minute. Push hard, with a depth of at least two inches for adults.

- **Rescue Breaths**: After every 30 compressions, deliver two rescue breaths by tilting the victim's head back, lifting their chin, and blowing air into their mouth to inflate their lungs.
- **Continue CPR**: Continue the cycle of 30 compressions and two breaths until medical help arrives or the victim begins to breathe on their own.

It is important for firefighters to be certified in CPR and keep their skills up to date through regular training. In many cases, CPR can sustain life long enough for paramedics to arrive and take over, making it a crucial skill for firefighters.

3. Basic First Aid Procedures

In addition to CPR, firefighters must be proficient in administering first aid for a variety of injuries. The ability to stop bleeding, immobilize broken bones, and treat burns or other trauma is often required in the field.

Some of the most common first aid procedures include:

- **Bleeding Control**: For severe bleeding, apply direct pressure to the wound with a clean cloth or bandage. Elevating the injury and using tourniquets when necessary can help control blood loss.
- **Treating Burns**: Firefighters frequently deal with burn victims. For minor burns, cooling the area with water and applying a sterile dressing can help alleviate pain and prevent infection. For more severe burns, it is essential to cover the area with a clean, dry cloth and seek immediate medical attention.
- **Fracture Immobilization**: In cases of broken bones, firefighters must stabilize the injury to prevent further damage. This can be done by applying a splint or using available materials like boards or rolled-up clothing to immobilize the injured area.
- **Shock Management**: Victims of trauma are often at risk of going into shock, a life-threatening condition where the body's organs do not receive enough oxygen. Signs of shock include pale, clammy skin, rapid breathing, and confusion. Firefighters should keep the victim warm, elevate their legs, and provide reassurance until medical help arrives.

4. First Aid for Other Common Emergencies

Firefighters are also trained to respond to a wide range of medical emergencies, including:

- **Choking**: Firefighters must be able to perform the Heimlich maneuver to clear a person's airway in cases of choking.
- **Seizures**: Keeping the person safe and preventing injury during a seizure is the first thing to do. Firefighters should help by protecting the victim's head and clearing the area of dangerous objects.
- **Poisoning**: If poisoning is suspected, firefighters must identify the substance and relay that information to medical personnel while ensuring the victim is safe and stabilized.

5. Application in Firefighter Exams

First aid and CPR are commonly tested in firefighter entrance exams. Candidates should expect to demonstrate their ability to assess medical situations, perform CPR correctly, and administer first aid for various injuries. These scenarios may involve written questions or practical assessments, where candidates are asked to demonstrate their skills on mannequins or simulated victims.

Preparing for these exams involves regular practice of CPR techniques, studying first aid procedures, and staying informed about updates to medical guidelines. Firefighters must also be able to perform these skills under pressure, as emergency situations are often chaotic and require quick, decisive action.

Chapter 10: EMT Basics and Terminology

Firefighters are often called upon to perform Emergency Medical Technician (EMT) duties in the field. As first responders, they must be prepared to deliver life-saving medical care before paramedics or other healthcare professionals arrive. This chapter introduces the key responsibilities of an EMT and provides an overview of the essential medical terminology that every firefighter needs to know, both for the firefighter entrance exam and for real-world emergency situations.

1. The Role of an EMT in Firefighting

In many fire departments, firefighters are cross-trained as EMTs to respond to a variety of medical emergencies. An EMT's primary responsibility is to provide immediate care to injured or ill individuals, stabilize their condition, and prepare them for transport to a medical facility. The ability to administer basic emergency medical care, such as CPR, wound management, and airway support, is crucial for firefighters.

Some common EMT responsibilities include:

- **Assessing the Scene**: Firefighters must quickly evaluate the scene for hazards, assess the severity of injuries, and determine the appropriate course of action to prevent further harm to victims and bystanders.
- **Providing Basic Life Support (BLS)**: BLS includes administering CPR, using an automated external defibrillator (AED), managing airways, and providing ventilations. EMTs must be proficient in performing these tasks under high-pressure conditions.
- **Immobilizing and Transporting Victims**: In cases of trauma, EMTs stabilize broken bones or injuries using splints and stretchers. They also ensure that victims are safely transported to medical facilities without worsening their condition.

- **Administering Oxygen**: Many emergencies require EMTs to deliver oxygen to patients suffering from breathing difficulties. EMTs must understand how to operate oxygen tanks and masks to provide proper care.
- **Recording Patient Information**: EMTs are responsible for documenting the patient's medical condition, symptoms, and the care provided at the scene.

Firefighter candidates must be familiar with the basic duties of an EMT, as this knowledge is often tested on entrance exams.

2. Essential Medical Terminology for Firefighters

Understanding medical terminology is a required skill for firefighters, particularly when communicating with other healthcare professionals. Clear and precise language ensures that firefighters can accurately describe a victim's condition and the treatment administered. This section covers essential medical terms that are commonly used in emergency medical care.

- **Cardiac and Respiratory Terms**:
    - **Cardiopulmonary**: Relating to the heart and lungs, such as in **cardiopulmonary resuscitation (CPR)**.
    - **Apnea**: The absence of breathing.
    - **Tachycardia**: A condition where the heart beats faster than normal (over 100 beats per minute).
    - **Bradycardia**: A condition where the heart beats slower than normal (fewer than 60 beats per minute).
    - **Dyspnea**: Difficulty or labored breathing.

- **Injury and Trauma Terms**:
    - **Laceration**: A deep cut or tear in the skin.
    - **Contusion**: A bruise caused by trauma, where blood vessels are damaged under the skin.
    - **Fracture**: A break in the bone.

- **Shock**: A life-threatening condition where the body's organs do not receive enough oxygen due to poor blood flow.

- **Common Medical Conditions**:
  - **Anaphylaxis**: A severe allergic reaction that can be life-threatening, requiring immediate treatment.
  - **Seizure**: Uncontrolled electrical activity in the brain, leading to convulsions or sudden body movements.
  - **Stroke (CVA)**: A cerebrovascular accident where blood flow to the brain is interrupted, causing brain damage.
  - **Myocardial Infarction (MI)**: A heart attack, occurring when blood flow to the heart muscle is blocked.

- **Treatment and Care Terms**:
  - **IV (Intravenous)**: A method of delivering fluids or medications directly into a vein.
  - **AED (Automated External Defibrillator)**: A device used to shock the heart back into a normal rhythm during cardiac arrest.
  - **Intubation**: Inserting a tube into the airway to assist with breathing.

3. Preparation for EMT and Medical Terminology Exams

Candidates preparing for the firefighter entrance exam should expect questions that test their knowledge of EMT basics and medical terminology. These questions may include scenario-based assessments where candidates are asked to identify symptoms, choose the correct treatment, or define medical terms.

Some tips for exam preparation include:

- **Study common medical conditions**: Be able to recognize and understand the symptoms of medical emergencies, such as heart attacks, strokes, and anaphylactic reactions.

- **Practice medical terminology**: Review terms related to anatomy, medical equipment, and emergency procedures. Understanding how to break down complex medical words into their roots, prefixes, and suffixes can help with unfamiliar terms.
- **Review EMT procedures**: Study basic life support techniques, including how to assess a patient, provide CPR, manage airways, and stabilize injuries. Practice these procedures through hands-on training or simulation exercises to build confidence.

4. The Importance of EMT Training for Firefighters

While firefighting involves the physical suppression of fires, the ability to deliver emergency medical care is equally important. Firefighters who are trained as EMTs provide support in situations where time is of the essence, and their medical skills can save lives. Many fire departments now require EMT certification as part of their training, emphasizing its importance within the role of a firefighter.

Beyond passing the exam, firefighters must continually maintain their EMT certifications and stay updated on new medical practices and technologies. Fire departments regularly offer refresher courses and advanced medical training to ensure that firefighters remain equipped to handle any medical emergency they encounter in the field.

Part E: Firefighting Techniques

Chapter 11: Search and Rescue Techniques

Search and rescue operations require a blend of strategy, skill, and knowledge to execute effectively. These operations are not only about finding and saving lives but also about ensuring the safety of the rescuers. The complexity of search and rescue missions varies greatly depending on the scenario, ranging from residential fires to large-scale natural disasters. Each type of mission demands a unique approach, tailored to the specific challenges it presents.

One of the foundational elements of successful search and rescue operations is a thorough understanding of building layouts and construction types. Firefighters must be able to quickly navigate through complex structures, identifying potential hazards and locating victims in need of assistance. This knowledge is complemented by the use of specialized equipment, such as thermal imaging cameras, which allow rescuers to see through smoke and darkness, pinpointing the heat signatures of trapped individuals.

Effective communication is another cornerstone of search and rescue efforts. Teams must be able to coordinate their movements and share information rapidly and accurately. This is facilitated by the use of handheld radios and other communication devices, which keep all members of the operation connected despite the challenging conditions they face.

Training plays a crucial role in preparing firefighters for the demands of search and rescue missions. Through rigorous exercises and simulations, candidates learn how to assess situations quickly, make decisions under pressure, and apply their skills in a variety of emergency scenarios. These training programs emphasize not only individual competence but also teamwork and coordination, as the success of search and rescue operations often hinges on the ability of the team to function as a cohesive unit.

In addition to technical skills and equipment, search and rescue operations also require a deep understanding of human behavior. Firefighters must be able to recognize signs of distress and panic in victims, approaching them in a manner that calms and reassures.

As we delve deeper into the specifics of search and rescue techniques, it's important to note the diversity of scenarios that firefighters may encounter. From residential fires to industrial accidents, each situation presents its own set of challenges and risks. The following sections will explore these scenarios in greater detail, highlighting the strategies and tools that can be employed to enhance safety and efficiency in search and rescue operations.

Navigating through smoke-filled environments demands not only courage but also a profound understanding of the behavior of smoke and fire. Firefighters use techniques such as the "left-hand search" or "right-hand search" method to systematically explore a building while maintaining a reference point, reducing the risk of disorientation. This method involves choosing a wall to keep one hand on while searching, ensuring that the team can trace their way back if needed. Additionally, the practice of "sounding the floor" ahead with tools ensures that firefighters avoid weakened structures that could lead to collapse.

Understanding the dynamics of fire behavior in different settings is crucial. For instance, in a high-rise fire, the stack effect can significantly influence smoke movement and fire spread, requiring different strategies compared to a single-story residential fire. Firefighters must be adept at reading these conditions, adjusting their search and rescue tactics to accommodate the unique challenges presented by each environment.

Victim extraction is where firefighters must decide the safest and most efficient method to remove victims from danger. Techniques vary from the simple "fireman's carry" to more complex procedures requiring specialized equipment like stokes baskets or rescue stretchers. In scenarios involving hazardous materials, rescuers may also need to employ decontamination processes as part of the extraction effort, ensuring the safety of both the victim and the rescue team.

The importance of ongoing training cannot be overstated, as it equips firefighters with the latest techniques and knowledge to handle the evolving nature of search and rescue operations. Scenario-based training, including live fire exercises, simulates real-life conditions, allowing firefighters to practice decision-making and problem-solving in a controlled but realistic environment. This hands-on experience is invaluable, reinforcing the theoretical knowledge acquired through classroom learning.

Moreover, the psychological support for both victims and firefighters plays a pivotal role in the aftermath of rescue operations. Providing immediate emotional support to victims can help mitigate the impact of trauma, while post-incident debriefings for rescue teams offer a space to process experiences, discuss what was learned, and address any emotional fallout. These debriefings are essential for maintaining the mental health and readiness of firefighters for future operations.

Scenario 1: Residential Fire - Trapped Family in a Second-Floor Bedroom

You arrive at the scene of a house fire that has rapidly engulfed the ground floor of a two-story residence. Flames are spreading through the kitchen and living area, cutting off the main staircase. A family of four is trapped in a second-floor bedroom. They are visible at the window, calling for help. The fire is advancing, and time is running out.

- **Decision Point**: Do you attempt a rescue operation through a second-story window, potentially placing firefighters at risk due to the lack of access to the primary fire exits, or do you begin containing the fire on the ground floor to prevent the flames from reaching the family upstairs?
- **Key Considerations**: If the fire spreads to the second floor before the rescue can be completed, the family's survival chances decrease dramatically. However, attempting to extinguish the fire before rescue may delay the response, and conditions could worsen quickly for the victims. Using a ladder for a direct rescue might reduce the risk to the victims but could expose firefighters to danger if the fire spreads.

Scenario 2: Commercial Building Fire - Confined Fire vs. Large Number of Occupants

You respond to a fire in a large commercial building. The fire is currently contained to a storage room on the ground floor, but there are reports of multiple occupants on the upper floors, some of whom are unable to evacuate due to the building's complex layout. The fire is not yet out of control but has the potential to spread rapidly due to the building's ventilation system.

- **Decision Point**: Do you focus your efforts on evacuating the occupants from the upper floors, risking the possibility of the fire spreading unchecked through the ventilation system, or do you prioritize extinguishing the fire in the storage room to prevent further spread, potentially leaving occupants in a hazardous situation longer?
- **Key Considerations**: If the fire is contained quickly, it could save time and reduce the overall danger to occupants. However, delaying the evacuation may leave individuals trapped in smoke-filled areas, and they could suffer from smoke inhalation or be cut off by the fire if it spreads faster than anticipated.

Scenario 3: Vehicle Collision and Fire - One Person Trapped Inside, Fuel Leak Present

You arrive at the scene of a vehicle collision involving two cars. One vehicle is on fire due to a ruptured fuel line, and flames are beginning to spread toward the engine compartment. Inside the second vehicle, a driver is conscious but trapped by the wreckage. The fire has not yet reached the second vehicle, but fuel is leaking near the collision site, creating the risk of an explosion.

- **Decision Point**: Do you prioritize rescuing the trapped driver from the second vehicle, risking that the fire may spread and endanger both the rescue team and the victim, or do you first attempt to control the fire and stop the fuel leak to prevent an explosion, potentially delaying the driver's rescue?

- **Key Considerations**: The fire could quickly escalate if it reaches the leaking fuel, leading to an explosion that would threaten everyone in the vicinity. However, delaying the rescue to contain the fire increases the risk of injury to the trapped driver, especially if fire or smoke reaches the second vehicle during the firefighting efforts.

Firefighter exams often include scenario-based questions to assess a candidate's judgment, decision-making skills, and ability to prioritize actions under pressure. These scenarios are designed to simulate real-life situations firefighters may encounter, requiring a blend of knowledge, practical skills, and critical thinking.

Here are examples of multiple-choice questions to illustrate how these situations are presented and evaluated.

1. You arrive at the scene of a residential fire and observe smoke coming from the windows of a two-story house. A bystander informs you that there may be a person trapped inside on the second floor. Given the intensity of the fire and the potential risk of structural collapse, how do you proceed?
[A] Immediately enter the building to search for the trapped person.
[B] Use a thermal imaging camera from outside to assess the fire's location and intensity before deciding on entry.
[C] Wait for backup before attempting any rescue efforts.
[D] Focus on containing the fire from the outside to prevent it from spreading to neighboring structures.

2. During a search and rescue operation in a commercial building fire, you and your team find an unconscious victim. The path you used to enter is now blocked by fire. Which alternative escape route do you choose?
[A] Attempt to extinguish the fire blocking your original path.
[B] Use a secondary stairwell on the opposite side of the building.
[C] Break through a nearby window and use a ladder to descend.
[D] Wait in place for the fire to be extinguished by other team members.

3. You are the first to respond to a wildfire threatening a residential area. You must quickly decide between beginning evacuation procedures for the homes in the path of the fire or trying to create a firebreak to slow the fire's progress. Which action do you prioritize?
[A] Start immediate evacuation of residents.
[B] Focus on creating a firebreak with available equipment.
[C] Call for additional resources before making a decision.
[D] Evacuate the most vulnerable residents first, then work on the firebreak.

4. Upon arriving at a car accident scene where a vehicle is on fire, you learn that there is a person trapped inside the car. The fire is rapidly spreading, and there's a risk of the vehicle exploding. How do you approach this situation?
[A] Use a fire extinguisher to fight the fire while others attempt to rescue the trapped person.
[B] Prioritize extracting the person immediately, using hydraulic rescue tools.
[C] Establish a safety perimeter and wait for the fire to be controlled before attempting rescue.
[D] Attempt to suppress the fire from a distance while assessing the victim's condition.

5. While conducting a search operation inside a burning warehouse, you come across a room filled with unknown chemicals that are starting to react to the heat, posing a risk of explosion. What is your immediate action?
[A] Continue the search, ignoring the chemicals.
[B] Evacuate the area and report the hazardous materials to the incident commander.
[C] Try to identify and contain the chemicals yourself.
[D] Use water to cool down the chemicals, reducing the risk of explosion.

Answer Explanations:

1. **Residential Fire - Person Trapped Inside**
    - **Correct Answer**: **[B] Use a thermal imaging camera from outside to assess the fire's location and intensity before deciding on entry.**

- **Explanation**: Rushing into a fire without assessing the situation can be dangerous for both the trapped individual and the firefighter. Using a thermal imaging camera allows firefighters to evaluate the intensity and spread of the fire and locate any potential hotspots. This method provides a safer way to determine whether entering the structure is feasible without risking collapse. Option A puts the firefighter in immediate danger, and Option D ignores the possibility of a life still being saved inside. Waiting for backup in Option C may delay rescue operations.

2. **Commercial Building Fire - Unconscious Victim**
   - **Correct Answer**: **[C] Break through a nearby window and use a ladder to descend.**
   - **Explanation**: When your original path is blocked by fire, it's essential to find an alternative exit quickly. Breaking through a nearby window and descending with a ladder provides an efficient means of escape. Option A (attempting to extinguish the fire) would waste precious time, and Option D (waiting for the fire to be extinguished) poses unnecessary risk. Option B may take longer and isn't guaranteed to be a safe escape route if the fire spreads rapidly.

3. **Wildfire - Residential Evacuation vs. Firebreak**
   - **Correct Answer**: **[A] Start immediate evacuation of residents.**
   - **Explanation**: In a wildfire scenario, the safety of residents is the top priority. Evacuating people from the path of the fire is essential to prevent loss of life. Option B (creating a firebreak) would be a secondary concern after ensuring all residents are safe. Option C (calling for resources) would delay action, while Option D (evacuating only the vulnerable first) may leave others in the fire's path.

4. **Car Accident with Fire - Trapped Person Inside**
   - **Correct Answer**: **[B] Prioritize extracting the person immediately, using hydraulic rescue tools.**

- **Explanation**: The immediate extraction of the trapped person is the highest priority. Using hydraulic rescue tools allows you to free the person quickly, before the fire spreads further. Option A (using a fire extinguisher) might slow the fire but doesn't address the trapped individual's immediate danger. Option C (establishing a safety perimeter) and Option D (suppressing the fire) both delay rescue efforts.

5. **Warehouse Fire - Hazardous Chemicals**
   - **Correct Answer**: **[B] Evacuate the area and report the hazardous materials to the incident commander.**
   - **Explanation**: When unknown chemicals are reacting to heat, it's essential to evacuate and notify the incident commander to avoid exacerbating the situation. Chemicals may react unpredictably in a fire, so trying to contain them yourself (Option C) or using water (Option D) could cause further reactions or explosions. Option A (ignoring the chemicals) is highly dangerous and unacceptable.

Chapter 12: Ventilation Techniques

Ventilation is a firefighting technique that involves the systematic removal of hot smoke, gases, and fire products from a burning structure to improve visibility, reduce temperatures, and slow the fire's growth. By introducing fresh air into the environment, firefighters can more effectively locate and extinguish the fire, while also creating safer conditions for both victims and firefighting personnel. The primary goal of ventilation is to control the fire's behavior, making it easier to manage and ultimately extinguish. This process requires a deep understanding of fire behavior, building construction, and the principles of heat and smoke movement.

There are two main types of ventilation techniques used by firefighters: vertical and horizontal. Vertical ventilation involves creating openings in the roof or higher sections of a building to allow heat, smoke, and gases to escape upwards, following their natural tendency to rise. This method is particularly effective in large or complex structures where smoke can accumulate rapidly, posing significant challenges to firefighting efforts and victim rescue operations. Horizontal ventilation, on the other hand, focuses on creating openings at the same level as the fire or smoke, such as doors, windows, or walls. This technique is often used in conjunction with vertical ventilation to maximize airflow and smoke clearance.

The decision on which ventilation technique to use depends on various factors, including the type of structure, the stage of the fire, and the prevailing wind conditions. Firefighters must assess the situation quickly and choose the most effective method to achieve ventilation without exacerbating the fire's spread. For instance, improper timing or location of ventilation openings can introduce fresh oxygen to the fire, leading to rapid fire growth and increased danger.

In addition to these primary methods, firefighters also employ mechanical ventilation, using fans and blowers to force smoke and hot gases out of the structure. Mechanical ventilation can be highly effective in certain scenarios, but it requires careful coordination and understanding of airflow patterns within the building.

Understanding the dynamics of fire and smoke movement is essential for effective ventilation. The principles of the fire triangle (oxygen, heat, and fuel) and the behavior of smoke and hot gases under different conditions guide firefighters in creating an effective ventilation strategy. For example, recognizing the signs of a backdraft or flashover is crucial in timing ventilation efforts to prevent these dangerous phenomena from occurring.

The use of tools and equipment in ventilation operations is another important aspect. Firefighters use a variety of hand and power tools to create openings for ventilation, including axes, chainsaws, and specialized cutting equipment. The choice of tools depends on the materials and construction of the building, as well as the urgency of the ventilation needs.

Firefighters undergo extensive training to understand the theory behind ventilation, learn the use of tools and equipment, and practice decision-making in simulated fire scenarios.

The impact of ventilation on firefighting operations cannot be overstated. Proper ventilation not only facilitates fire suppression efforts but also significantly improves the chances of survival for trapped occupants. By reducing the heat and toxic smoke levels, firefighters can gain quicker access to the fire source and victims, enhancing the overall effectiveness of the firefighting operation.

As we continue to explore the intricacies of ventilation techniques, it's important to consider the advancements in technology and firefighting strategies that influence current practices. Innovations in equipment, materials, and building design all play a role in shaping the approaches to ventilation, requiring firefighters to stay informed and adaptable. The next section will delve further into the specific methods and considerations for effective ventilation in a variety of fire scenarios, highlighting the challenges and solutions encountered by firefighters in their efforts to protect lives and property.

As firefighters assess the environment and determine the best approach for ventilation, they must also consider the impact of environmental factors such as wind direction and speed. Wind-driven fires present a unique challenge, as strong winds can quickly carry heat, smoke, and fire to other parts of the building or even to nearby structures. In such cases, strategic placement of ventilation openings is crucial to harness the wind's force in a way that aids firefighting efforts without accelerating the fire's spread. Firefighters may use wind to their advantage by creating openings on the leeward side of the structure, allowing the wind to push smoke and heat out of the building.

Another consideration in ventilation operations is the preservation of structural integrity. When performing vertical ventilation, firefighters must be aware of load-bearing elements and avoid compromising the building's stability. This requires knowledge of building construction and the ability to quickly identify safe areas for creating openings. The risks of structural collapse are a constant concern, and the safety of firefighting personnel and any occupants is always the top priority.

Mechanical ventilation, while effective, must be used judiciously. The introduction of positive pressure ventilation fans can significantly improve air quality and visibility inside a structure, but if not properly coordinated, it can also push fire, smoke, and gases into uninvolved areas, creating new hazards. Firefighters must carefully monitor conditions and adjust their ventilation tactics as the situation evolves, ensuring that mechanical ventilation supports their overall strategy for fire suppression and rescue.

The role of training and continuous learning cannot be understated in the context of ventilation and firefighting as a whole. Firefighting is a dynamic field, with new challenges emerging as building designs and materials evolve. Ongoing education in fire science, building construction, and ventilation techniques is essential for firefighters to stay ahead of these changes and refine their strategies for effective fire suppression and rescue.

Moreover, the psychological aspect of firefighting, particularly during high-stakes ventilation operations, requires resilience and mental preparedness. Firefighters must be able to remain calm and focused amidst the chaos of a fire scene, making rapid decisions based on their training, experience, and the conditions at hand. This mental fortitude is cultivated through rigorous training exercises that simulate the stress and unpredictability of real fire scenarios.

In addition to physical and mental preparation, effective communication and teamwork are pivotal during ventilation operations. Coordination between team members ensures that ventilation efforts are synchronized with search and rescue, fire suppression, and other activities on the fireground. The ability to communicate clearly and efficiently under pressure is a skill that firefighters develop through practice and experience.

Advancements in technology also play a significant role in enhancing ventilation techniques. Innovations such as drone technology for aerial reconnaissance and thermal imaging for identifying heat sources and ventilation needs are becoming increasingly integral to firefighting operations. These tools provide valuable information that helps firefighters make informed decisions about where and how to ventilate, ultimately improving the effectiveness and safety of their efforts.

The complexities of ventilation in firefighting underscore the importance of a comprehensive approach that combines knowledge, skill, and technology. By understanding the principles of fire behavior and building construction, employing the appropriate techniques and tools, and working as a cohesive team, firefighters can effectively manage smoke, heat, and fire, protecting lives and property. As the field of firefighting continues to evolve, so too will the strategies and technologies for ventilation, requiring a commitment to lifelong learning and adaptation from those who bravely undertake this work.

Scenario 1: Ventilating a Two-Story Residential Structure Fire

You and your team respond to a fire in a two-story residential building. The fire started on the first floor in the living room, but smoke is rapidly filling the second floor, trapping an occupant in a bedroom. The fire has not yet reached the second floor, but visibility and air quality are deteriorating.

- **Decision Point**: Do you choose vertical ventilation by cutting a hole in the roof to allow the smoke to escape upward, or do you use horizontal ventilation by breaking windows on the second floor to clear the smoke at the same level as the trapped occupant?

- **Key Considerations**: Vertical ventilation would allow heat and smoke to escape upward, creating better conditions inside for search and rescue. However, it requires time and manpower. Horizontal ventilation is quicker but may risk feeding the fire with oxygen if not timed properly. Choosing the right technique depends on the fire's progression and the time available to rescue the occupant.

Scenario 2: Wind-Driven Fire in a High-Rise Building

A fire has broken out on the 10th floor of a high-rise apartment building. Strong winds are pushing the smoke and flames toward the upper floors. Several occupants are trapped in their apartments. The fire is rapidly spreading due to the wind conditions, and the heat is intensifying on the windward side of the building.

- **Decision Point**: Do you create a ventilation opening on the windward side, risking that the wind will intensify the fire's spread, or do you create an opening on the leeward side (the side away from the wind) to allow the wind to push smoke out of the building?
- **Key Considerations**: Ventilation on the windward side could worsen the fire's spread by introducing more oxygen to the flames, while leeward-side ventilation allows the wind to aid in smoke removal. However, the structural stability of the building must also be considered before performing any ventilation.

Scenario 3: Industrial Fire with Hazardous Chemicals and Smoke Accumulation

You are called to an industrial warehouse fire where chemicals are stored. The fire is generating a significant amount of toxic smoke, and the structure is beginning to fill with thick, hazardous fumes. Several firefighters inside are reporting difficulty with visibility and deteriorating air quality, but the fire is not yet under control.

- **Decision Point**: Do you prioritize mechanical ventilation using fans to clear the smoke quickly, or do you employ vertical ventilation by creating roof openings to allow the heat and smoke to escape naturally?

- **Key Considerations**: Mechanical ventilation is faster and could provide immediate relief to the firefighters inside, improving air quality. However, without careful coordination, it could inadvertently push toxic smoke to other parts of the warehouse or nearby structures. Vertical ventilation is slower but may provide a more controlled smoke release. The decision will depend on the urgency of the air quality issue and the complexity of the building's layout.

Here are examples of multiple-choice questions to illustrate how these situations are presented and evaluated.

1. During an industrial complex fire, you are tasked with ventilating a structure filled with flammable gases. The risk of explosion is high. Which ventilation method do you employ to minimize the risk of igniting the gases?
[A] Horizontal ventilation to quickly remove gases at their level
[B] Vertical ventilation to allow gases to rise and disperse naturally
[C] Mechanical ventilation using fans to force out the gases
[D] No ventilation, to avoid introducing oxygen that could fuel the fire

2. You're faced with a basement fire in a residential home. The fire is generating a lot of smoke, making it difficult to see and breathe. What is your first step in ventilating this area to improve conditions for entry?
[A] Break windows at ground level to introduce fresh air into the basement
[B] Use positive pressure ventilation at the top of the stairs to push smoke out
[C] Open the basement door to allow smoke to escape naturally
[D] Create an opening in the floor above the fire to use vertical ventilation

3. In a high-rise fire, you need to prevent smoke from spreading to upper floors where evacuations are still underway. Which ventilation strategy is most effective in this scenario?
[A] Horizontal ventilation on floors with active fire to remove smoke
[B] Vertical ventilation through stairwells to draw smoke upwards and out
[C] Sealing off the fire floor and using mechanical ventilation to pressurize adjacent floors
[D] Opening windows on the fire-affected floor to allow smoke to escape

4. After extinguishing a fire in a commercial building, you notice that smoke is still lingering in the air, posing a health risk. What is the best approach to clear the smoke efficiently?
[A] Use fans to create a cross-flow of air, pushing smoke out of open windows and doors
[B] Employ vertical ventilation techniques to let the hot smoke rise and exit through the roof
[C] Increase mechanical ventilation to circulate fresh air and remove smoke
[D] Wait for the smoke to dissipate naturally, avoiding further disturbance to the structure

5. You arrive at a scene where a fire has been mostly contained, but there's a risk of re-ignition due to hot spots and smoldering areas. How do you ventilate the structure to prevent flare-ups while avoiding excessive air movement that could reignite the fire?
[A] Strategically place fans to gently move air across hot spots without increasing oxygen flow directly to the fire
[B] Use vertical ventilation to draw heat and smoke upwards, carefully monitoring for any signs of flare-ups
[C] Open all available windows and doors to increase natural airflow and cool down hot spots gradually
[D] Apply minimal mechanical ventilation at low speed to avoid disturbing smoldering areas while removing hot air

Answer Explanations:

1. **Industrial Complex Fire - Ventilating a Structure Filled with Flammable Gases**
   - **Correct Answer**: **[B] Vertical ventilation to allow gases to rise and disperse naturally.**
   - **Explanation**: In a scenario involving flammable gases, vertical ventilation is the safest option because it allows the gases to rise and disperse through openings at higher levels, reducing the concentration of gas in the building.

Horizontal ventilation (Option A) or mechanical ventilation (Option C) could cause the gases to ignite by pushing them around, while not ventilating (Option D) increases the danger by allowing gases to accumulate.

2. **Basement Fire in a Residential Home - Initial Ventilation Steps**
   - **Correct Answer**: **[B] Use positive pressure ventilation at the top of the stairs to push smoke out.**
   - **Explanation**: Positive pressure ventilation (PPV) creates a controlled flow of air that forces smoke out of the basement, improving visibility and air quality for entry. Breaking windows (Option A) could lead to improper ventilation, and opening the basement door (Option C) may worsen conditions by allowing more smoke to accumulate. Vertical ventilation (Option D) might work in larger fires but could be slower to implement in a basement scenario.

3. **High-Rise Fire - Preventing Smoke Spread to Upper Floors**
   - **Correct Answer**: **[C] Sealing off the fire floor and using mechanical ventilation to pressurize adjacent floors.**
   - **Explanation**: Sealing the fire floor and using mechanical ventilation to pressurize nearby floors prevents smoke from spreading upward, protecting occupants still evacuating. Horizontal ventilation (Option A) or vertical ventilation (Option B) could move the smoke through stairwells or other openings, and opening windows (Option D) may make the smoke spread uncontrollably.
   -

4. **Clearing Smoke After a Fire in a Commercial Building**
   - **Correct Answer**: **[A] Use fans to create a cross-flow of air, pushing smoke out of open windows and doors.**
   - **Explanation**: Creating a cross-flow with fans is the most efficient way to remove lingering smoke, as it clears the air quickly by pushing it out through

windows and doors. Vertical ventilation (Option B) is slower and may not be as effective after the fire is extinguished, while relying on mechanical ventilation (Option C) may take longer and waiting for the smoke to dissipate naturally (Option D) could endanger health.

5. **Preventing Re-Ignition After a Fire - Hot Spots and Smoldering Areas**
   - **Correct Answer**: **[D] Apply minimal mechanical ventilation at low speed to avoid disturbing smoldering areas while removing hot air.**
   - **Explanation**: Using low-speed mechanical ventilation ensures that hot air is removed without disturbing smoldering areas, minimizing the risk of re-ignition. Using fans to move air across hot spots (Option A) or using vertical ventilation (Option B) may increase the airflow too much, providing oxygen that could reignite the fire. Opening all windows and doors (Option C) would create too much airflow and could lead to flare-ups.

Chapter 13: Water Supply Management

Managing water supply require firefighters to have a comprehensive understanding of various water sources and how to effectively use hydrants during emergencies. Water, being the primary extinguishing agent for fires, necessitates that firefighters are adept in quickly locating and accessing water supplies, ensuring there is enough water pressure, and understanding the dynamics of water flow to combat fires efficiently. The process begins with identifying potential water sources, which can range from municipal water systems, static sources like ponds or lakes, to portable water tanks in areas where hydrants are not readily available. Each source presents its own set of challenges and requires specific strategies to utilize effectively.

Hydrants serve as the most common and reliable source of water for firefighting in urban and suburban areas. Knowledge of the hydrant system, including the location, type, and operation is essential for firefighters. Hydrants are color-coded to indicate water flow capacity, a crucial factor in planning firefighting operations. Firefighters must be familiar with the color codes and understand how to interpret them to ensure they are tapping into a hydrant with sufficient flow rate for the fire at hand. Additionally, regular maintenance and inspection of hydrants help ensure they are operational when needed. This includes checking for leaks, making sure caps are not seized due to corrosion, and confirming the hydrant is accessible and unobstructed.

The technique of drafting water from static sources is employed in rural or remote areas where hydrants are not available. This involves using a fire engine's pump to draw water from a pond, lake, or portable tank. Drafting requires specific equipment such as hard suction hoses and strainers to prevent debris from entering the pump and damaging it.

Water shuttle operations are another method used in areas lacking a direct water supply. This involves transporting water from a source to the fire scene using tanker trucks or tenders. The efficiency of a water shuttle operation depends on the coordination between the vehicles, the distance to the water source, and the ability to quickly load and unload water. Planning and practicing water shuttle operations are crucial for rural firefighting, where this method may be the primary means of water supply.

Understanding the principles of water pressure and flow is fundamental to managing water supply in firefighting. Firefighters must calculate the required water flow to extinguish a fire, taking into account factors such as the size of the fire, the type of materials burning, and the method of attack. This calculation informs decisions on the number of hoses to deploy, the size of nozzles, and the pump pressure required to deliver water effectively to the fire. Pump operators play a key role in this process, adjusting the pump to achieve the desired pressure and flow rate based on the length and diameter of hoses, elevation changes, and other factors affecting water movement.

In summary, managing water supply in firefighting operations involves a multifaceted approach that includes understanding and accessing various water sources, effectively using hydrants, drafting from static sources, coordinating water shuttle operations, and mastering the principles of water pressure and flow. Firefighters must be proficient in these areas to ensure they can provide a sufficient and continuous water supply to combat fires effectively. The next section will continue to explore the intricacies of water supply management, focusing on advanced techniques and strategies for optimizing water use during firefighting operations.

Advanced techniques in water supply management for firefighting operations involve not only the physical aspects of accessing and deploying water but also strategic planning and technological advancements that enhance efficiency and effectiveness. One such technique is the use of water relay systems, which can extend the reach of water supply lines over long distances. This method involves multiple pumpers placed at intervals, each boosting the water pressure to the next, allowing water to be transported over greater distances without significant loss of pressure. This is particularly useful in large-scale incidents or in areas where the water source is far from the fire scene.

The integration of technology into water supply management has also seen significant advancements. Geographic Information Systems (GIS) and mobile applications are now used to map hydrant locations, water mains, and even static water sources. This technology provides firefighters with real-time information on the nearest water sources, their flow capacities, and any relevant operational issues, such as out-of-service hydrants. Such tools not only save valuable time during an emergency but also assist in pre-incident planning and simulation exercises.

Another aspect of water supply management is the establishment of water supply task forces or strike teams in regions prone to large-scale fires or where water supply challenges are prevalent. These specialized units are equipped with high-capacity pumpers, water tenders, and portable water tanks, and are trained in rapid deployment and operation of complex water supply systems. Their expertise allows for a more coordinated and effective response to incidents where water access is a significant challenge.

In addition to technological and strategic advancements, the principles of conservation and efficiency in water use cannot be overlooked. Firefighting foam systems, for instance, play a crucial role in enhancing the effectiveness of water in fire suppression. Foam additives reduce the surface tension of water, allowing it to spread and penetrate more effectively, thereby reducing the overall volume of water needed to extinguish a fire. This is particularly important in environmental conservation and in situations where water supply is limited.

Training and continuous education are the backbone of effective water supply management in firefighting. Firefighters must stay abreast of the latest techniques, technologies, and best practices through regular training sessions, workshops, and participation in simulation exercises. These educational opportunities not only improve individual skills but also foster teamwork and coordination among firefighters, which are essential for successful water supply operations.

The role of community involvement and education in water supply management should also be emphasized. Public awareness campaigns on the importance of hydrant accessibility, for example, can encourage residents to keep hydrants clear of obstructions and report any damages or operational issues. Such community engagement enhances the overall efficiency of the firefighting efforts and ensures a more resilient response to fire emergencies.

As firefighting operations continue to evolve, so too will the strategies and technologies for water supply management. The future may see further advancements in water delivery systems, such as autonomous or remotely operated water tenders, and even greater use of data analytics and artificial intelligence to predict water needs and optimize supply routes. The commitment to innovation, training, and community partnership will remain central to overcoming the challenges of water supply management in firefighting, ensuring that firefighters have the necessary resources to protect lives and property effectively.

Part F: Fire Prevention and Safety

Chapter 14: Fire Codes and Regulations

Fire codes and regulations are fundamental in preventing fires, protecting lives, and ensuring property safety. These codes serve as a framework of rules and guidelines that govern the design, construction, and maintenance of buildings and public spaces to minimize fire hazards. Firefighters, building inspectors, architects, and property owners must work within these regulations to reduce the risks associated with fire outbreaks and ensure rapid, safe emergency responses when fires occur.

1. Understanding Fire Codes

Fire codes are developed and maintained by various agencies, most notably the National Fire Protection Association (NFPA) and local government bodies. The NFPA develops codes and standards that are adopted widely across the United States, including the widely recognized **NFPA 1: Fire Code** and **NFPA 101: Life Safety Code**. These standards cover all aspects of fire prevention and safety, from the design of fire exits and alarm systems to the materials used in construction. Additionally, states and municipalities often adopt their own codes based on these standards, adapting them to address local risks and specific building types.

Key elements of fire codes include:

- **Fire Exits**: Regulations ensure that buildings have a sufficient number of marked and unobstructed exits, allowing occupants to evacuate safely during an emergency.
- **Sprinkler Systems**: Fire suppression systems, particularly automatic sprinklers, are required in many buildings to prevent fires from spreading before firefighters arrive.
- **Alarm Systems**: Properly installed and maintained fire alarm systems ensure that occupants and emergency personnel are alerted promptly when a fire occurs.

- **Fire Walls and Doors**: Special construction materials and designs (like fire-rated doors and walls) are required to contain fires within specific sections of a building, limiting the spread of flames and smoke.

2. Building and Maintenance Regulations

In addition to fire codes, building codes provide detailed instructions for construction practices that minimize fire hazards. **Building materials**, for example, must meet certain flame resistance ratings to reduce the speed at which fire spreads. Modern building codes also emphasize **structural fire resistance**, requiring that certain components—like beams and load-bearing walls—can withstand high temperatures without collapsing.

Maintenance regulations are equally important. Fire protection equipment (e.g., sprinklers, extinguishers, and alarms) must undergo regular inspections to ensure they are in working order. Building owners and property managers are responsible for conducting these inspections, often on a quarterly or annual basis, depending on local laws.

3. Occupancy-Specific Regulations

Different types of buildings face unique fire risks, and fire codes are tailored to these needs. For example:

- **Commercial Buildings**: Offices, stores, and industrial facilities must comply with stringent regulations regarding sprinkler systems, alarm systems, and accessible exits. High-occupancy buildings like malls and theaters must have additional fire safety measures in place.
- **Residential Buildings**: Multi-family dwellings, such as apartment complexes, must meet fire safety regulations regarding exits, fire escapes, and the availability of fire extinguishers. Smoke detectors and sprinkler systems are also mandatory in certain residential settings.
- **Educational and Healthcare Facilities**: These settings require special attention due to the vulnerable nature of occupants. In schools and hospitals, fire codes emphasize the importance of evacuation routes, fire drills, and smoke

control systems to ensure a quick and orderly evacuation of students, patients, and staff.

4. Fire Safety Inspections and Enforcement

Fire codes are enforced through regular **fire safety inspections** conducted by local fire departments or building safety inspectors. These inspections ensure that property owners and managers comply with fire safety laws and make any necessary corrections. Inspections typically involve:

- **Checking fire extinguishers**, alarm systems, and sprinkler systems to ensure they are in working order.
- **Reviewing exit routes** to confirm they are accessible, unobstructed, and clearly marked.
- **Verifying the maintenance** of fire doors, smoke control systems, and fire walls.
- **Ensuring compliance** with occupancy-specific requirements, such as the proper storage of flammable materials in industrial settings.

Failure to comply with fire codes can result in fines, shutdowns, or other legal actions. In severe cases where public safety is at risk, a building can be condemned until the fire safety violations are corrected.

5. Fire Prevention Programs and Public Education

In addition to fire codes, fire departments and local governments run **public fire prevention programs** aimed at educating the public about fire safety. These programs often include fire drills, safety demonstrations, and the distribution of educational materials. Firefighters play a key role in these efforts, helping to raise awareness of the importance of fire alarms, safe evacuation routes, and proper handling of flammable materials.

Public education campaigns often target high-risk groups such as children, the elderly, and occupants of high-density housing areas. By teaching basic fire prevention strategies—such as "stop, drop, and roll" and safe cooking practices—these programs significantly reduce the likelihood of fire-related injuries and fatalities.

6. Adapting to Changing Fire Risks

As urban landscapes evolve and building technologies advance, fire codes must be continuously updated to address new fire risks. For example, the increased use of **solar panels, lithium-ion batteries**, and **energy-efficient building materials** introduces new challenges for firefighting. Modern codes are beginning to reflect these changes by requiring updated fire suppression systems and strategies for handling emerging hazards like electrical fires.

Additionally, as cities grow more densely populated, new fire safety regulations focus on **high-rise buildings** and **complex urban environments**. Fire departments must adjust their strategies and equipment to manage fires in taller buildings, often relying on advanced technology such as **thermal imaging cameras** and **drone reconnaissance** to ensure fire codes are followed in these challenging settings.

Fire codes and regulations form the backbone of fire safety, guiding the construction and maintenance of buildings to prevent fires and mitigate their damage when they occur. By understanding and enforcing these regulations, firefighters and safety inspectors protect lives, reduce property damage, and ensure the efficient operation of emergency services. Continuous education, inspections, and updates to fire codes ensure that as fire risks evolve, so too do the strategies to combat them. Through rigorous enforcement and public education, fire departments and local authorities ensure that the risk of fire is minimized, and that communities are prepared for any fire-related emergencies.

Chapter 15: Hazardous Materials Safety

Hazardous materials, often referred to as HazMat, include a wide range of chemicals and substances that can be flammable, explosive, toxic, or corrosive. The first step in managing these materials safely is proper identification. Firefighters must be familiar with the National Fire Protection Association (NFPA) 704 Diamond, a standard marking system that uses color-coded fields and numbers to indicate the type of hazard and its severity. This system helps first responders quickly assess the risks associated with a particular substance at a glance.

Handling hazardous materials requires specialized training and equipment. Firefighters must wear appropriate personal protective equipment (PPE) to shield themselves from harmful exposure. This gear can include respirators, gloves, chemical-resistant suits, and boots. The type of PPE required depends on the specific HazMat involved and the nature of the exposure. For example, dealing with airborne toxic substances might necessitate the use of self-contained breathing apparatus (SCBA) to prevent inhalation.

Safety procedures for managing hazardous materials incidents are comprehensive and multifaceted. They include establishing a safety perimeter to limit exposure to the HazMat, evacuating civilians from the affected area, and implementing decontamination procedures for both the responders and potentially contaminated individuals. Decontamination is a crucial step in mitigating the spread of hazardous substances and involves removing or neutralizing contaminants from people, equipment, and the environment.

Firefighters also rely on the Incident Command System (ICS) to manage HazMat incidents effectively. The ICS is a standardized approach to the command, control, and coordination of emergency response, providing a common hierarchy within which responders from multiple agencies can be effective. Under the ICS, specific roles are designated for managing different aspects of the incident, including safety, operations, planning, and logistics.

Understanding the behavior of hazardous materials in fire conditions is another essential aspect of safe handling. Some substances may react violently when exposed to heat or water, creating additional hazards such as toxic fumes, explosions, or chemical burns. Firefighters must be knowledgeable about the chemical properties of hazardous materials and the appropriate extinguishing agents and techniques to use in various situations.

To effectively manage hazardous materials incidents, firefighters must also be proficient in the use of specialized equipment designed for HazMat situations. This includes chemical detection devices that can identify and quantify the presence of toxic substances, as well as containment tools such as overpack drums and non-sparking tools to safely secure and transport hazardous materials. The goal is to minimize the risk of contamination or further release of hazardous substances into the environment.

Coordination with hazardous materials response teams (HMRTs) is crucial during significant HazMat incidents. These specialized teams possess advanced training and equipment to handle a wide range of hazardous substances and scenarios. Collaboration between firefighters and HMRTs ensures that incidents are managed with the highest level of expertise and safety.

Regular drills that simulate various hazardous materials scenarios help responders practice identification, containment, decontamination, and coordination procedures. These exercises also allow firefighters to familiarize themselves with the equipment and protective gear required for HazMat responses, ensuring they are ready to act quickly and efficiently during an actual incident.

Firefighters and HazMat teams often engage in community outreach to inform the public about the risks associated with hazardous materials and the importance of proper storage, handling, and disposal. Educating the public can prevent accidents and reduce the likelihood of hazardous materials incidents.

Finally, continuous learning and adherence to updated guidelines and regulations are essential for firefighters dealing with hazardous materials. As new substances are developed and existing materials are better understood, protocols for safe handling, containment, and neutralization evolve. Staying informed about the latest developments in hazardous materials safety ensures that firefighters are always prepared to respond effectively to HazMat incidents, protecting themselves, the public, and the environment from the dangers these materials pose.

Chapter 16: Fireground Safety Procedures

Fireground operations present a unique set of challenges and dangers, making safety procedures paramount to protect both firefighters and civilians. One of the cornerstone principles of fireground safety is the establishment of a command structure, which ensures that all operations are coordinated under a single commander. This structure is established with the purpose of maintaining communication and organization, preventing freelancing, and ensuring that all personnel are accounted for at all times. The Incident Command System (ICS) provides a standardized approach to the command, control, and coordination of emergency response efforts, facilitating a more efficient and effective operation.

Another safety procedure is the use of Personal Protective Equipment (PPE). Firefighters must be equipped with the appropriate gear, including turnout gear, self-contained breathing apparatus (SCBA), gloves, and helmets, to protect against heat, flames, smoke, and other hazardous materials they might encounter. Regular checks and maintenance of PPE are essential to ensure that all equipment functions correctly when needed.

The use of accountability tags or systems helps track all firefighters and personnel on the scene, monitoring their entry and exit from hazardous areas. This system is crucial for initiating a rapid intervention team (RIT) response if a firefighter becomes lost, trapped, or incapacitated.

Ensuring effective communication is another fundamental aspect of fireground safety. Handheld radios, tactical channels, and clear communication protocols are necessary to coordinate efforts, call for assistance, and relay information about the evolving situation. Miscommunication can lead to delays, errors, and potentially catastrophic outcomes, emphasizing the need for clear and concise exchanges.

Fireground operations also require a thorough understanding of building construction and behavior under fire conditions. Knowledge of different building types, materials, and structural integrity can significantly impact firefighting tactics and strategies. Firefighters must be able to quickly assess structures for potential collapse hazards, escape routes, and the best approach for controlling the fire.

Scene safety is further enhanced by securing a water supply and establishing hose lines efficiently. A reliable water source is crucial for firefighting efforts, and hose lines must be deployed in a manner that allows for quick advancement and retreat, minimizing the risk of entanglement or obstruction. Firefighters must also be adept at recognizing signs of backdraft, flashover, and other rapid fire progression phenomena to avoid being caught in sudden, explosive fire growth.

In addition to these procedures, continuous situational awareness is imperative for all individuals on the fireground. Firefighters must constantly evaluate their environment for changes in fire behavior, structural stability, and potential hazards, adjusting their tactics accordingly. This vigilance helps in making informed decisions and taking appropriate actions to mitigate risks and protect lives.

Training and preparedness are the foundations of fireground safety. Regular drills, simulations, and continued education on fire behavior, equipment use, and emergency medical care ensure that firefighters are ready to respond effectively to any situation. This ongoing commitment to learning and improvement is crucial for maintaining high safety standards and enhancing the overall effectiveness of fireground operations.

Collaboration with other emergency services, such as police and emergency medical services (EMS), is also essential for a coordinated response to fire incidents. Establishing clear roles, communication channels, and joint protocols facilitates a unified approach to incident management, ensuring that all aspects of the emergency are addressed comprehensively.

The use of technology and advanced tools, such as drones for aerial surveillance and thermal imaging cameras, has become increasingly important in fireground operations. These technologies provide valuable insights into the fire's extent, hotspots, and potential victims, aiding in strategic planning and resource allocation. However, reliance on technology should not replace fundamental firefighting skills and instincts, which remain essential for navigating the unpredictable nature of fire incidents.

In summary, safety procedures on the fireground encompass a broad range of practices, from personal protection and accountability to strategic planning and situational awareness. These measures are designed to protect firefighters and civilians alike, ensuring that fire incidents are managed with the highest degree of professionalism and care. As fireground operations continue to evolve with advancements in technology and understanding of fire behavior, so too will the approaches to ensuring safety and efficiency in responding to these emergencies.

Effective scene lighting is crucial for nighttime or low-visibility operations, ensuring that firefighters can navigate safely and efficiently. Portable lights, truck-mounted floodlights, and even helmet-mounted lights play a significant role in illuminating the operational area, reducing the risk of accidents and enhancing the ability to locate victims or hazards quickly. Proper illumination supports safe movement around the fireground and aids in the identification of structural weaknesses or dangerous conditions that may not be visible under normal circumstances.

The establishment of control zones – hot, warm, and cold – is a strategic safety measure that segregates the fireground into areas based on hazard levels. The hot zone is where the immediate fire-related activities occur, and only personnel with essential roles and proper PPE are allowed entry. The warm zone serves as a buffer, where tasks such as staging and equipment preparation take place, and the cold zone is designated for command posts, rehabilitation, and medical treatment areas, ensuring a safe distance from the immediate dangers of the fire scene.

Rehabilitation protocols are essential for maintaining firefighter health and safety during prolonged incidents. These protocols include medical evaluations, hydration, rest, and recovery periods, protecting firefighters from overexertion, heat stress, or other health issues. The rehab area, typically located in the cold zone, provides a space for firefighters to recover and prepare for redeployment, ensuring they are physically and mentally ready to continue operations.

The implementation of rapid intervention teams (RIT) or firefighter assist and search teams (FAST) is a proactive approach to firefighter safety, providing dedicated personnel ready to respond to emergencies involving firefighters on the scene. These teams are equipped and trained to perform rescues of trapped or injured firefighters, demonstrating a commitment to the well-being of fire service personnel. Their presence ensures that a systematic and immediate response is available should a firefighter encounter distress or danger during operations.

Fireground operations must also include strategies for environmental protection, including containment of runoff water to prevent pollution and the spread of hazardous materials. Firefighters must be aware of the environmental impact of firefighting foam and other extinguishing agents, employing best practices to minimize harm to waterways, soil, and surrounding ecosystems. This consideration is crucial for preserving public health and the environment, highlighting the broader responsibilities of firefighters beyond the immediate task of extinguishing fires.

Lastly, post-incident analysis and debriefings are invaluable for enhancing future fireground safety and operations. These sessions provide opportunities to review actions taken, identify areas for improvement, and reinforce successful strategies. Learning from each incident is a key component of continuous improvement, ensuring that lessons are integrated into training and operational planning. This reflective practice supports the development of more effective and safer fireground tactics, benefiting both firefighters and the communities they serve.

Initial Fire Cause Determination

Firefighters first on the scene play a pivotal role in this process, as their observations can significantly influence the direction of the subsequent investigation. The ability to distinguish between signs of arson and accidental fires hinges on recognizing specific indicators present at the scene. Arson, intentionally set fires, often leaves behind clues such as multiple points of origin, the use of accelerants, and unusual burn patterns. Accelerants, substances used to initiate or accelerate the spread of fire, can be detected through the presence of strong odors reminiscent of gasoline or kerosene, and their residue may be found on surfaces or absorbed in materials at the scene. Burn patterns, particularly those that suggest a fire spread unnaturally fast or originated in multiple, unconnected locations, can also indicate arson.

Conversely, accidental fires usually have a single point of origin and can be traced back to common sources such as electrical faults, unattended candles, cooking equipment, or smoking materials. Electrical fires might leave behind charred outlets, melted wires, or blown fuses, indicating where the fire started. Kitchen fires often originate from stovetops or ovens and may be accompanied by grease buildup, which can fuel the fire. Smoking-related fires might start in areas where smoking materials are typically used, such as living rooms or bedrooms, and are often found near upholstered furniture or bedding.

The presence of safety devices such as smoke detectors and fire suppression systems, or the lack thereof, can also provide insights into the nature of the fire. In arson cases, these systems may have been tampered with or disabled to allow the fire to spread unchecked. In accidental fires, the failure of these systems to activate may point to a lack of maintenance or malfunction.

Firefighters and investigators also look for signs of forced entry or exit, which can be indicative of arson if it appears that someone entered the property to set the fire and then left in haste. However, signs of forced entry can also be the result of rescue attempts or the actions of the occupants trying to escape the fire.

The analysis of debris and the examination of fire dynamics are essential for understanding how the fire started and spread. Samples of burned materials can be sent to laboratories for chemical analysis to detect the presence of accelerants, while the study of fire dynamics can reveal whether the fire's behavior was consistent with natural progression or influenced by external factors.

In the initial stages of fire cause determination, it is crucial to approach the scene with an open mind, considering all possible causes and meticulously documenting evidence. Photographs, notes, and physical samples help piece together the sequence of events leading up to the fire. Collaboration with other emergency response teams, including law enforcement and forensic specialists, improves the accuracy and thoroughness of the investigation.

Training and experience are invaluable in accurately determining the cause of a fire. Firefighters and investigators must stay informed about the latest techniques and technologies in fire analysis, as well as common patterns associated with different types of fires. Continuous education and practical exercises help develop the skills necessary to identify subtle clues and differentiate between arson and accidental fires effectively.

Ultimately, the goal of initial fire cause determination is to establish a factual basis for the fire's origin, which can inform prevention strategies, support criminal investigations, and facilitate insurance and legal processes. By carefully analyzing the evidence and applying their knowledge of fire behavior, firefighters and investigators can uncover the truth behind the flames, ensuring justice and safety for the community.

Chain of Custody Protocols

Maintaining the **chain of custody** is crucial in ensuring that evidence collected at the scene of a fire is preserved and remains untampered, which is essential for any subsequent legal or investigative processes. The chain of custody refers to the chronological documentation or paper trail, showing the seizure, custody, control, transfer, analysis, and disposition of physical or electronic evidence. It is a process that begins the moment evidence is collected from the scene and continues until it is presented in court or otherwise disposed of.

1. Evidence Collection: The first step in maintaining a proper chain of custody starts with the collection of evidence. Firefighters and investigators must wear gloves and use appropriate tools to collect evidence to prevent contamination. Each piece of evidence should be placed in a separate, properly labeled evidence bag or container. The label should include a description of the item, the location where it was found, the date and time of collection, and the name of the person who collected it.

2. Documentation: Every action taken with the evidence must be documented. This includes the initial collection, any examination or analysis that occurs, and the storage and transfer of evidence. Documentation should be detailed, including dates, times, and the names of individuals involved in each step. This meticulous record-keeping is essential for establishing a continuous chain of custody.

3. Storage and Transportation: Once collected, evidence must be stored and transported in a manner that protects it from damage, contamination, or loss. Evidence should be stored in a secure, climate-controlled environment to prevent degradation. When evidence needs to be transported, it should be done so in a secure manner, with documentation of each person who handles it during the transfer.

4. Access Control: Only authorized personnel should have access to evidence, and their interaction with the evidence should be documented. Access logs can help track who has handled or viewed the evidence and for what purpose.

5. Analysis and Examination: When evidence undergoes analysis or examination, the process should be conducted by qualified professionals who follow standardized procedures. Each step of the analysis should be documented, including the results and any conclusions drawn from the evidence. This documentation becomes part of the chain of custody and is crucial for interpreting the evidence's relevance to the investigation.

6. Legal Transfer: If evidence is required for legal proceedings, its transfer to the court or legal representatives must be carefully managed and documented. The transfer should include a detailed record of the evidence being handed over, including a signature from the receiving party acknowledging receipt.

7. Disposition: The final disposition of evidence, whether it is returned to the owner, destroyed, or retained for further analysis, must also be documented. The decision on the disposition of evidence should be made in accordance with legal requirements and departmental policies.

Role of Fire Investigators

Fire investigators play a crucial role in the aftermath of a fire, working meticulously to determine the cause and origin of the incident. Their work is pivotal in identifying whether a fire was accidental or intentional, which can have significant legal and insurance implications. Firefighters assist these investigators by helping to preserve the integrity of the fire scene, ensuring that valuable evidence is not destroyed or contaminated during firefighting operations. The collaboration between firefighters and fire investigators begins with the understanding that every action taken at the scene can impact the subsequent investigation. From the moment firefighters arrive on the scene, they are trained to observe and mentally note the condition of the scene, including the location of doors, windows, and the presence of any potential fire accelerants. This initial assessment is crucial for investigators who will later rely on these observations to piece together the events leading up to the fire.

Firefighters are also trained to minimize water and chemical damage when possible, understanding that these elements can wash away or dilute potential evidence. In cases where the use of water or firefighting foam is unavoidable, firefighters aim to use these substances judiciously, focusing on suppressing the fire while preserving the scene's integrity. The approach to extinguishing the fire, including the entry points used and the areas targeted, is conducted with an awareness of potential evidence areas. Firefighters avoid unnecessary disturbance to the fire debris, knowing that the patterns of burn and the distribution of debris can tell the story of how the fire spread.

Once the fire is under control, firefighters may establish a security perimeter to prevent unauthorized access to the scene. Firefighters and fire investigators work together to document the scene comprehensively, using photographs, sketches, and notes. This documentation process includes capturing the scene's condition before any salvage or overhaul operations are conducted. Salvage operations, aimed at protecting valuable property from water, smoke, and fire damage, are carried out with a keen awareness of the need to preserve evidence. Overhaul operations, which involve searching for hidden fire pockets and ensuring the fire is completely extinguished, are performed carefully to avoid disturbing evidence.

Firefighters assist investigators by identifying and preserving items that could serve as evidence, such as remnants of electrical appliances, pieces of fabric that might indicate the presence of accelerants, or any unusual objects that could be related to the fire's cause. These items are flagged for the investigators, who will later examine them more closely. The role of firefighters extends to providing firsthand accounts of the fire scene upon their arrival and during firefighting operations. These accounts can offer valuable insights into the fire's behavior and progression.

In cases where arson is suspected, firefighters are particularly vigilant in preserving the scene for law enforcement and fire investigators. They understand that their actions and observations can be crucial in building a case against suspects. The collaborative effort between firefighters and fire investigators is a testament to the multidisciplinary approach required in the field of fire and emergency services. This partnership ensures that the cause of the fire is accurately determined, contributing to justice, safety, and prevention efforts.

Part G: Cognitive and Test-Taking Skills

Chapter 17: Test-Taking Strategies

Mastering the art of test-taking is crucial for candidates preparing for the Firefighter Entrance Exam. This chapter delves into effective strategies that can significantly enhance your performance on the exam. The focus here is on understanding the types of questions you will encounter and how to manage your time efficiently during the exam. A strategic approach to these aspects can make a substantial difference in your overall score and your chances of success.

Firstly, familiarize yourself with the format of the exam. Knowing whether you will be facing multiple-choice questions, true/false statements, or scenario-based inquiries is fundamental. Each type requires a slightly different strategy. For multiple-choice questions, it's essential to read each question carefully and eliminate the obviously incorrect answers first. This process of elimination can increase your chances of selecting the correct answer from the remaining options. Additionally, pay attention to the wording of the questions and answers. Sometimes, the phrasing can provide clues to the correct answer or eliminate potential choices.

For true/false questions, the key is to look for absolute terms such as "always" or "never." Statements with these words are more likely to be false, as they leave no room for exceptions. However, be cautious and use your knowledge and reasoning skills to assess each statement thoroughly before deciding.

When it comes to scenario-based questions, which are common in firefighter exams to assess situational judgment and decision-making skills, the best approach is to visualize yourself in the scenario. Consider the principles and training relevant to the situation described and choose the response that aligns with standard firefighting practices and safety protocols.

Begin by quickly scanning the entire exam to gauge the types and lengths of questions. This overview will help you allocate your time more effectively. For instance, if you notice a mix of multiple-choice and essay questions, you might decide to tackle the multiple-choice section first to secure those points, as they often require less time per question than essays or short-answer questions.

One effective time management technique is to answer the questions you know first, skipping over the ones you're unsure about. This ensures that you don't waste precious time pondering difficult questions while missing the opportunity to answer several easier ones. After completing all the questions you're confident about, return to the ones you skipped and apply the process of elimination or your best-educated guess.

Remember, your first instinct is often correct, so avoid changing your answers unless you find clear evidence or recall information that convinces you your initial choice was wrong. Trusting your preparation and instincts can be a valuable asset during the exam.

Another crucial strategy is to practice pacing yourself with timed practice tests. These simulations can help you become accustomed to the pressure of the exam environment and improve your ability to manage time effectively. By identifying areas where you tend to spend too much time, you can adjust your approach and improve your speed and accuracy.

In addition to these strategies, maintaining a positive mindset and managing exam stress are essential components of successful test-taking. Techniques such as deep breathing, visualization, and positive self-talk can help keep anxiety at bay and allow you to focus on the task at hand. Remember, preparation is key. The more familiar you are with the material and the exam format, the more confident and calm you will be on exam day.

As we continue to explore test-taking strategies, we will delve into more specific techniques and considerations that can further enhance your exam performance. These include understanding the importance of reading comprehension in the context of the exam, strategies for tackling complex problem-solving questions, and tips for ensuring your physical and mental well-being in the days leading up to the exam.

Understanding the importance of reading comprehension cannot be overstated, especially in an exam setting where every word counts. The ability to quickly read, comprehend, and analyze information is crucial, particularly for scenario-based and problem-solving questions. Enhance your reading skills by practicing with materials similar to those you'll encounter on the exam. Focus on summarizing paragraphs in your own words and identifying key points and details. This practice will not only improve your speed but also your ability to extract essential information from complex texts.

When tackling complex problem-solving questions, start by breaking down the question into manageable parts. Identify what is being asked and separate it from the supplementary information. Sometimes, questions are designed to test your ability to sift through data and focus on what's relevant. Drawing diagrams or jotting down key points can help visualize the problem and guide you to a solution. If calculations are required, double-check your work for simple mistakes, such as misplaced decimal points or incorrect units of measurement.

In the days leading up to the exam, taking care of your physical and mental well-being is as important as your academic preparation. Ensure you're getting enough sleep, eating nutritious foods, and engaging in physical activity to keep your body and mind in optimal condition. Avoid cramming the night before the exam, as this can lead to burnout and increased anxiety. Instead, review your notes and practice materials in a relaxed manner, focusing on your areas of strength to boost your confidence.

During the exam, if you encounter a question that seems insurmountable, don't let it derail your focus. Mark it and move on, allowing yourself time to address it later. Sometimes, answers or ideas come to you while you're working on other questions. This approach prevents you from getting stuck and helps maintain a steady pace throughout the exam.

Lastly, be mindful of the physical exam environment. Wear comfortable clothing and bring all necessary supplies, including several pencils, an eraser, and a watch to keep track of time if clocks aren't visible in the exam room. Arrive early to the testing site to acclimate to the surroundings and reduce any last-minute stress.

By integrating these strategies into your exam preparation, you're not just preparing to pass; you're setting yourself up for success. The firefighter entrance exam is a comprehensive test of your knowledge, skills, and ability to perform under pressure. With diligent preparation, a strategic approach to tackling different types of questions, and a focus on your well-being, you can approach the exam with confidence and achieve your goal of becoming a firefighter. Remember, each question is an opportunity to demonstrate your readiness for the challenges of firefighting, so approach the exam with determination and focus.

Chapter 18: Memorization Techniques

Firefighters are expected to recall vast amounts of information, from fire codes and safety protocols to equipment details and emergency procedures, often under intense conditions. However, memorizing this extensive body of knowledge can be daunting, especially when preparing for rigorous exams or facing real-life emergencies. Here are some proven strategies to help with the retention and recall of vital firefighting information.

1. **Method of Loci (Memory Palace Technique)**

   This technique involves associating information with specific familiar locations. For example, to remember different classes of fires and the appropriate extinguishing methods, imagine placing each class in a different room of your house. Visualize unique cues in each room that relate to the characteristics of each fire class, making the details easier to retrieve when needed.

2. **Mnemonics**
   Mnemonics help in retaining information through abbreviations, acronyms, or catchy phrases. For instance, the acronym "PASS" aids in recalling fire extinguisher steps: Pull the pin, Aim at the base, Squeeze the lever, and Sweep the nozzle. Creating vivid or humorous mental images that represent the information further enhances retention, as unusual and sensory-rich images are easier to remember.

3. **Chunking**
   This method involves breaking down large amounts of information into smaller, manageable units or "chunks." For numerical sequences, such as emergency codes, group numbers into smaller sets to simplify memorization. For example, instead of remembering ten digits as a whole, divide them into three smaller groups.

4. **Spaced Repetition and Active Recall**

   Repetition over spaced intervals helps strengthen memory. By reviewing material over increasing time intervals—known as spaced repetition—you reinforce long-term retention. Combine this with active recall, where you test yourself without referring to notes, to improve both recall speed and accuracy.

5. **Visualization**
   Visualization involves creating a mental image of the information. This technique is especially useful for memorizing sequences, procedures, or equipment layouts. By mentally picturing each step or component, you build a mental map that can be easily recalled in exams or field scenarios.

6. **Storytelling**
   Humans naturally remember stories better than isolated facts. By transforming the information into a story, you make it more engaging and memorable. For example, to remember the steps of a hazardous material response, create a narrative where each step is a plot point in the story, making the sequence easier to recall.

7. **Interleaved Practice**

   Rather than focusing on one topic at a time (blocked practice), mix different subjects or material types together. For instance, when studying fire codes, intersperse questions on equipment and procedures. This technique challenges the brain to adapt to varied information, enhancing both memory retention and application skills.

8. **Teaching Others**

   Explaining concepts or procedures to others reinforces your own understanding. Teaching a study group or even explaining a concept to a friend compels you to clarify your thoughts, identify knowledge gaps, and strengthen retention through repetition.

9. **Healthy Lifestyle**

   Cognitive function and memory are significantly influenced by lifestyle. Prioritizing adequate sleep, regular exercise, and balanced nutrition boosts your capacity to concentrate and retain information. Getting sufficient rest, especially before study sessions, enhances focus, while physical activity is linked to improved memory.

Chapter 19: Visualization and Spatial Skills

In firefighting, visualization and spatial orientation help firefighters navigate complex environments and respond effectively to emergencies. These skills are also tested in firefighter exams through scenario-based questions, where candidates must mentally picture scenes, assess spatial layouts, and make quick decisions based on the situation.

1. The Importance of Spatial Awareness

Spatial awareness refers to the ability to recognize and understand the relationships between objects and the environment, especially in terms of position, distance, and movement. For firefighters, spatial awareness is crucial for several reasons:

- **Navigating unfamiliar environments**: In smoke-filled buildings with low visibility, firefighters must rely on their mental mapping of the space to move through the structure, locate fire sources, and rescue trapped individuals.
- **Positioning firefighting equipment**: Understanding spatial dynamics helps firefighters place hoses, ladders, and other equipment in optimal positions to control the fire while maintaining safe exits.
- **Avoiding hazards**: Firefighters must be able to anticipate the behavior of fire and how it interacts with structural elements, allowing them to avoid collapsing ceilings, weakened floors, and other dangers.

In firefighter exams, spatial awareness is often tested through scenario-based questions that simulate real-world fire environments. Candidates are expected to demonstrate their ability to visualize spaces, predict fire behavior, and navigate complex layouts safely.

2. Visualization Techniques

Visualization is the mental process of forming images or concepts of scenarios in one's mind, which can be used to plan actions or solve problems. For firefighters, visualization is key to anticipating what they cannot immediately see, such as hidden flames or smoke in adjacent rooms.

Developing strong visualization skills allows firefighters to:

- **Plan escape routes**: Even before entering a burning building, firefighters need to visualize potential exit routes in case conditions worsen inside.
- **Anticipate fire spread**: By visualizing how heat, smoke, and flames might move through a building, firefighters can strategically position themselves to cut off fire paths and protect occupants.
- **Understand building layouts**: Visualizing the structure's layout helps firefighters determine where to search for victims, how to ventilate the space effectively, and where fire suppression efforts should be concentrated.

In exam scenarios, visualization helps candidates mentally prepare for various challenges, such as where fire might spread next, or how to move through a building without visibility. Strong visualization skills also assist in mechanical reasoning questions, where candidates must visualize how firefighting equipment operates in different settings.

3. Developing Spatial Orientation Skills

Spatial orientation refers to the ability to maintain a sense of direction and location within a space, even when faced with confusing or disorienting environments, such as during a fire or when surrounded by smoke. For firefighters, mastering spatial orientation is needed for:

- **Keeping track of exits**: Knowing the quickest way out of a burning building, even when visibility is low, ensures that firefighters can evacuate safely and assist others in doing so.
- **Locating key structures**: In the chaos of a fire, identifying load-bearing walls, stairwells, and other important structural elements helps firefighters make decisions about where to go and what to avoid.

- **Coordinating with team members**: When firefighters are dispersed throughout a building, maintaining spatial orientation ensures that they know where their team is positioned and how to navigate back to them if needed.

Firefighter exams often challenge candidates with spatial orientation tasks, such as determining the best route through a building or estimating the location of unseen fire hazards based on provided information. Developing these skills requires practice with 3D spatial exercises, floor plans, and interactive simulations.

4. Techniques for Improving Visualization and Spatial Orientation

To enhance these skills, firefighter candidates can engage in various exercises designed to sharpen their ability to mentally map and navigate environments.

Techniques include:

- **Practicing with floor plans**: Study blueprints or floor plans of different building types (e.g., residential, commercial, industrial) and visualize navigating through them under fire conditions. Mentally assess where potential hazards might occur and how to approach fire suppression or rescue.
- **Using virtual reality simulations**: Many fire departments now offer virtual reality (VR) training, allowing firefighters to practice navigation, fire behavior analysis, and equipment usage in a controlled, yet realistic, environment.
- **Playing spatial reasoning games**: Games and exercises that challenge spatial reasoning, such as puzzles or 3D visualization apps, can help candidates think critically about space and improve their mental mapping abilities.
- **Visualizing fire scenes**: After reviewing fire incident reports or case studies, visualize the fire scene as described, thinking through how the fire might have spread and how you would navigate the space.

5. Application in Firefighter Exams

Many firefighter exams include spatial reasoning and scenario-based questions to test a candidate's ability to mentally visualize environments and make quick decisions under pressure. For example, candidates may be presented with a 2D layout of a building and asked to identify the best route to rescue victims or choose the most effective position to set up equipment.

In these scenarios, strong visualization and spatial orientation skills allow candidates to:

- **Identify hazards**: Quickly recognize areas where fire is likely to spread or where the building might be structurally compromised.
- **Plan actions**: Map out a logical approach to managing the fire, such as determining where to place ventilation openings or how to prioritize victim rescues.
- **Maintain safety**: Assess the space to ensure that both firefighters and occupants are safe throughout the operation, and adjust plans based on new developments.

Chapter 20: Reading and Writing Skills

Improving reading comprehension and written expression is fundamental for firefighter candidates preparing for their entrance exams. These skills are not only crucial for the written part of the exam but are also essential for effective communication in the field. To enhance reading comprehension, start by actively reading a variety of materials related to firefighting, such as textbooks, technical manuals, and safety protocols. This will not only build your vocabulary but also familiarize you with the types of texts you might encounter on the exam and in your future career. When reading, try to summarize paragraphs in your own words, focusing on the main ideas and supporting details. This practice helps in retaining information and understanding complex concepts more deeply.

For written expression, clarity and conciseness are key. Begin by writing summaries of articles or chapters you've read, paying close attention to accurately conveying the original message in a more condensed form. Practice writing incident reports based on hypothetical scenarios, ensuring that you include all relevant information in a logical and organized manner. This exercise will improve your ability to communicate effectively under pressure, a skill that is invaluable in emergency situations.

Another effective strategy is to engage in discussions about firefighting techniques, equipment, or tactics with peers or mentors. This not only enhances your understanding of the material but also allows you to practice articulating your thoughts clearly and concisely. Additionally, participating in study groups can provide opportunities to receive feedback on your written work, which is crucial for identifying areas for improvement.

Incorporating vocabulary exercises into your study routine can also significantly benefit your reading comprehension and written expression. Create flashcards with technical terms and definitions, and regularly test yourself. This not only aids in memorization but also ensures you are comfortable using specialized terminology in your writing.

To further develop your reading skills, practice identifying the author's purpose and tone in various texts. This will help you become more adept at quickly understanding and analyzing information, a skill that is particularly useful for answering exam questions efficiently. Additionally, try to predict content and ask questions about the text before you begin reading. This active engagement with the material can improve your comprehension and retention rates.

For those struggling with grammar and punctuation, consider using online resources or workbooks specifically designed to address common writing issues. Regular practice with these tools can help you avoid mistakes that could detract from the clarity of your written communication.

Remember, improving reading comprehension and written expression takes time and dedication. By incorporating these strategies into your study routine, you can enhance your performance on the firefighter entrance exam and prepare yourself for a successful career in firefighting.

To bolster your written expression further, practice drafting responses to essay questions that mirror the style and substance of those found on the firefighter entrance exam. This will not only hone your ability to construct well-organized and coherent arguments but also familiarize you with the exam's format and expectations. Focus on developing a clear thesis statement for each essay and support your arguments with specific examples, particularly those related to firefighting scenarios and protocols.

After writing, take the time to review your work for any spelling, grammar, or punctuation errors, as well as for clarity and flow. This attention to detail can significantly impact the quality of your writing and your score on the written portion of the exam. To aid in this process, consider using digital tools that can help identify and correct common errors, but don't rely solely on them. Developing your own ability to spot and correct mistakes is invaluable.

In addition to individual study, participating in writing workshops or classes can provide structured guidance and feedback from experienced instructors. These settings often offer the chance to engage in peer review sessions, where you can gain insights into your writing strengths and areas for improvement from the perspective of your fellow candidates. Such collaborative learning environments can be incredibly beneficial for developing both reading comprehension and written expression skills.

For reading comprehension, diversifying your reading materials can also be beneficial. Don't limit yourself to texts strictly related to firefighting. Engaging with a wide range of genres, including fiction, non-fiction, and journalistic writing, can expose you to different writing styles and broaden your understanding of language and structure. This exposure can enhance your ability to quickly adapt to and understand the variety of texts you might encounter on the exam.

Lastly, consider integrating technology into your preparation. Numerous apps and online platforms offer practice tests, flashcards, and interactive games designed to improve vocabulary, grammar, and reading comprehension. These resources can provide a more dynamic and engaging way to study, making your preparation for the firefighter entrance exam both effective and enjoyable.

By dedicating time to develop your reading comprehension and written expression skills through these varied methods, you will not only prepare yourself for success on the firefighter entrance exam but also lay a solid foundation for effective communication in your future career as a firefighter. Regular practice, combined with a strategic approach to learning and improvement, will ensure that you are well-equipped to tackle the challenges of the exam and beyond.

Part H: Math, Mechanical, Spatial Aptitude

Chapter 21: Basic Math and Hydraulics

Firefighting often requires quick, precise calculations in high-pressure situations. Understanding basic math and hydraulic principles is crucial for tasks such as determining water flow, pressure, and hose line management during fire suppression efforts. This chapter aims to equip candidates with foundational math and hydraulic skills, alongside practical calculations, to help them succeed in exams and real-world scenarios.

1. The Role of Math in Firefighting

Firefighters frequently use math to make decisions about water requirements, hose setup, and equipment use. Here's how these calculations apply in practice:

- **Estimating Water Needs**: To determine water needs, firefighters estimate the size and severity of a fire. For example, if a fire requires 250 gallons per minute (GPM) and is expected to burn for 30 minutes, they calculate the water needed as follows:

$$250\, GPM \times 30\, minutes = 7500 gallons$$

This estimate helps firefighters prepare adequate resources to control the fire.

- **Pressure Calculations**: Proper pressure is essential to deliver water effectively. If a hose requires a nozzle pressure of 100 psi (pounds per square inch) and the pump is located 200 feet away, friction loss needs to be accounted for to maintain pressure. For example, if the friction loss is estimated at 10 psi per 100 feet:

$$10\, psi\ per\ 100\, ft \times 2 = 20\, psi\ loss$$

Therefore, the pump must deliver at least 120 psi to maintain the needed pressure at the nozzle.

- **Hose Length and Friction Loss**: Friction loss depends on hose length and diameter. For a 200-foot, 1.5-inch diameter hose delivering 125 GPM, friction loss can be calculated using charts or standard formulas. For instance:

$$Friction\ Loss = \left(\frac{125\ GPM}{100}\right)^2 \times 2 = 31.25\ psi$$

Firefighters can use this calculation to choose the most effective hose setup.

2. Essential Math Concepts for Firefighters

The following math concepts are essential for firefighting, with practical calculations included for clarity:

- **Basic Arithmetic**: Quick arithmetic helps firefighters manage resources, like calculating water reserves. If a tank holds 1000 gallons and is dispensing at 100 GPM, firefighters calculate remaining water as follows:

$$1000\ gallons - (100\ GPM \times 5\ minutes) = 500\ gallons\ left$$

- **Fractions and Percentages**: Fractions and percentages apply to mixing agents like foam. If a 3% foam solution is required in a 100-gallon tank:

$$100\ gallons \times 0.03 = 3\ gallons\ of\ foam\ concentrate$$

This helps ensure accurate mixing for effective fire suppression.

- **Area and Volume Calculations**: Knowing area and volume aids in estimating water needs. For example, if a room measures 20 feet by 15 feet with an 8-foot ceiling, calculate the room's volume:

$$20 \times 15 \times 8 = 2400\ cubic\ feet$$

With this information, firefighters can estimate the water needed based on the room's size.

- **Ratios and Proportions**: Ratios help with adjustments. If a fire truck pump produces 300 GPM at 100 psi, doubling the pressure may decrease the flow due to equipment limits. Firefighters use ratios to maintain the right balance between pressure and flow.

3. Introduction to Hydraulics

Hydraulic concepts help firefighters control water effectively, with calculations supporting their understanding:

- **Friction Loss**: Knowing friction loss helps adjust water pressure. For a 2.5-inch hose delivering 250 GPM over 300 feet, calculate friction loss with standard charts or estimates:

$$Friction\ Loss = \frac{250^2}{100} \times 1.5\ \times 3\ = = 56.25\ psi$$

This helps determine the pressure needed at the pump to ensure effective water delivery.

- **Water Pressure**: Water pressure affects nozzle performance. For a nozzle needing 100 psi, if it's 50 feet above the pump, account for elevation loss (0.434 psi per foot):

$$50\ feet \times 0.434 = 21.7\ psi$$

The pump must deliver 121.7 psi to overcome elevation and reach the nozzle effectively.

- **Elevation and Gravity**: Firefighters consider gravity's effect on pressure when pumping water uphill or downhill. If pumping water 40 feet uphill, calculate pressure loss as follows:

$$40 \times 0.434 = 17.36\ psi\ loss$$

Adjusting for this loss helps maintain water flow and prevent equipment strain.

4. Practical Applications of Hydraulics

Hydraulic principles play a major role in pump operations, hose line management, and nozzle settings:

- **Pump Operations**: Firefighters must calculate the correct pump pressure to deliver water efficiently. For example, if friction loss totals 40 psi over a 200-foot hose, and the nozzle requires 80 psi, the pump must deliver:

$$80 + 40 = 120\,psi$$

This ensures adequate pressure at the nozzle to suppress the fire.

- **Hose Line Management**: Choosing the correct hose diameter and length minimizes friction loss and maintains pressure. For a longer hose (e.g., 300 feet) with 1.75-inch diameter, delivering 150 GPM, friction loss might be calculated as:

$$Friction\ Loss = \left(\frac{150}{100}\right)^2 \times 3\ \approx 67.5\,psi$$

Knowing this, firefighters select hoses that balance length with flow requirements.

- **Nozzle Flow Rates**: Different nozzles have specific flow rates. For example, a nozzle set at 100 GPM at 50 psi can be adjusted based on fire needs. If more water is needed, firefighters may increase the pressure or switch to a higher GPM nozzle.

5. Key Formulas for Firefighter Math and Hydraulics

To prepare for both the firefighter exam and fieldwork, candidates should memorize several key formulas, including:

- **Friction Loss Formula**: Used to calculate how much water pressure is lost as it flows through hoses:

$$FL = C \times \frac{Q^2 \times L}{d^5}$$

Where FL is friction loss, C is a coefficient based on hose type, Q is flow rate, L is hose length, and d is hose diameter.

- **Flow Rate Formula**: Calculates how much water is being delivered at the nozzle:

$$Q = \frac{29.7 \times d^2\sqrt{P}}{GPM}$$

Where Q is the flow rate, d is the nozzle diameter, and P is the pressure.

- **Elevation Pressure**: Accounts for changes in pressure based on building height or elevation:

$$EP=0.434\times h$$

Where EP is elevation pressure and h is the height of the building in feet.

Chapter 22: Mechanical Reasoning Skills

Firefighters must not only be adept at using various mechanical devices, but they must also understand the underlying principles that govern how these tools work. This chapter focuses on developing mechanical aptitude and reasoning skills, which are essential for both the firefighter entrance exam and day-to-day firefighting operations.

1. What is Mechanical Reasoning?

Mechanical reasoning involves the ability to understand and apply the principles of mechanics, including force, motion, leverage, and basic physics, to real-world problems. For firefighters, mechanical reasoning is necessary for tasks such as operating hydraulic rescue tools, ladders, fire engines, and pumps. It also includes an understanding of the mechanical systems in buildings—like doors, windows, elevators, and HVAC systems—that firefighters might need to manipulate or disable in emergency situations.

Firefighter exams often include mechanical reasoning questions designed to test a candidate's ability to understand how things work, predict mechanical outcomes, and solve practical problems that arise during firefighting operations.

2. Key Mechanical Concepts for Firefighters

Firefighters need to be familiar with several key mechanical concepts to excel in both exams and real-life scenarios. These include:

- **Force and Leverage**: Understanding how force is applied and how leverage can be used to amplify force is essential when using tools such as axes, pry bars, and hydraulic spreaders. Leverage allows firefighters to exert a greater force with less effort, making it easier to open doors, breach walls, or extricate trapped individuals from vehicles.
- **Friction and Resistance**: When operating mechanical equipment, friction plays a major role in determining how efficiently tools perform. Whether it's managing the friction between a hose and the ground or understanding how friction in

moving parts can wear down equipment, firefighters must anticipate and adjust for resistance in all their operations.

- **Mechanical Advantage**: This concept refers to the amplification of input force to achieve a greater output force. Many firefighting tools, such as pulleys, wedges, and hydraulic jacks, operate on the principle of mechanical advantage, allowing firefighters to move heavy objects, open jammed doors, or lift debris with less effort.
- **Simple Machines**: Simple machines such as levers, pulleys, inclined planes, and screws are the building blocks of many firefighting tools. Firefighters must understand how these simple machines work to use more complex devices effectively.

3. Firefighting Equipment and Mechanical Operations

The operation of firefighting tools and equipment requires a solid understanding of mechanical principles. Some examples include:

- **Hydraulic Rescue Tools (Jaws of Life)**: These powerful tools rely on hydraulic pressure to cut through metal and spread apart wreckage, such as in vehicle extrications. Firefighters must understand how hydraulic systems work, how to control pressure, and how to safely position the tools for maximum efficiency.
- **Ladders and Aerial Devices**: Deploying ladders and aerial platforms requires knowledge of balance, stability, and load-bearing capacity. Firefighters must calculate the correct angle for ladder placement to ensure stability and safety, especially when extended to great heights or under weight-bearing conditions.
- **Fire Pumps and Hoses**: The mechanics of operating fire pumps involve understanding how pressure and flow rates are controlled to ensure water is delivered effectively to the nozzle. Firefighters must calculate and adjust pump pressure while managing friction loss in hoses to maintain an adequate water supply.
- **Ventilation Fans**: Mechanical ventilation fans are used to clear smoke from buildings. Firefighters must know how to position these fans effectively and

understand airflow principles to ensure that smoke is directed out of the structure without unintentionally feeding oxygen to the fire.

4. Mechanical Reasoning in Firefighter Exams

In firefighter entrance exams, mechanical reasoning questions are designed to assess a candidate's ability to apply mechanical principles in practical scenarios. These questions may involve:

- **Identifying simple machines** (e.g., pulleys, levers) and understanding how they can be used to gain mechanical advantage.
- **Predicting mechanical outcomes**, such as how adjusting the angle of a ladder affects its stability, or how using a longer lever increases the force applied to an object.
- **Problem-solving scenarios** that require mechanical reasoning, such as determining the most effective tool to open a jammed door or calculating the best way to extricate a person from a wrecked vehicle.

Candidates are expected to demonstrate a working knowledge of mechanical principles and apply them to various situations that they might encounter on the fireground.

5. Developing Mechanical Aptitude

Mechanical aptitude can be developed through hands-on experience with firefighting tools and equipment, as well as through training that emphasizes the physics and mechanics behind these devices. To improve mechanical reasoning, firefighter candidates can:

- **Practice using tools**: Become familiar with the mechanical operations of hydraulic spreaders, ladders, pumps, and other firefighting equipment. Learning how to operate these tools effectively will enhance a firefighter's ability to problem-solve in real-time.
- **Study mechanical principles**: Review the basic physics concepts behind force, motion, and leverage. Understanding these ideas in theory helps firefighters anticipate how equipment will behave in different conditions.

- **Participate in mechanical reasoning exercises**: These exercises often involve using simple machines like pulleys and levers, as well as solving mechanical puzzles that require thinking and the application of mechanical concepts.

Part I: Teamwork and Stress Management

Chapter 23: Psychology in Emergencies

Emergencies such as fires, vehicle accidents, and hazardous material incidents can provoke intense emotional responses from everyone involved, which can impact decision-making, communication, and overall effectiveness during firefighting operations. This chapter delves into the psychological impacts of emergencies on firefighters and victims, exploring typical behavioral responses and strategies for managing psychological stress during these high-pressure situations.

1. Understanding Human Behavior in Emergencies

During emergencies, human behavior can range from calm and rational to chaotic and irrational. Understanding these patterns of behavior helps firefighters manage chaotic scenes and assist victims effectively. Some common responses to emergencies include:

- **Panic**: When individuals experience fear, their natural fight-or-flight response can trigger panic. This can result in erratic behavior, poor decision-making, or an inability to follow instructions. Firefighters must be prepared to deal with panicked individuals who may not behave logically during rescue operations.
- **Freeze Response**: In some cases, individuals may freeze in the face of danger, becoming immobile and unable to act. This is a psychological response to overwhelming fear. Firefighters need to recognize when a person is in this state and take appropriate measures to guide or assist them to safety.
- **Denial**: Some victims may not fully comprehend the severity of the situation, leading to denial or refusal to evacuate. Firefighters must be skilled in communicating the urgency of the situation without exacerbating stress or fear.
- **Herd Behavior**: In large crowds, individuals often follow others, even when doing so may not be the safest option. Firefighters must manage crowd control to prevent harmful behaviors, such as stampedes or blockages at exits, which can lead to additional injuries or fatalities.

Recognizing these behavioral responses helps firefighters adapt their communication strategies and manage chaotic environments more effectively. For example, providing clear, direct instructions can help mitigate panic and confusion in emergency situations.

2. Psychological Stress in Firefighters

Firefighters experience high levels of stress during emergency operations, and understanding how to manage this stress is crucial for both individual well-being and operational effectiveness. The intense nature of firefighting, which often involves life-threatening situations, requires not only physical resilience but also strong psychological fortitude.

Common sources of stress for firefighters include:

- **Life-and-death decisions**: Firefighters are often placed in situations where their decisions can mean the difference between life and death, both for victims and for themselves.
- **Physical danger**: Exposure to fire, hazardous materials, and structural collapses presents a constant threat to safety, increasing stress levels.
- **Emotional strain**: Witnessing death, injury, and the suffering of victims can take an emotional toll on firefighters over time, leading to conditions like post-traumatic stress disorder (PTSD) if not properly managed.

To handle these stressors, firefighters can develop techniques to manage their psychological health and maintain mental clarity during emergencies.

These include:

- **Breathing and mindfulness techniques**: Practicing controlled breathing and mindfulness helps firefighters stay calm and focused in the midst of chaos. These techniques allow them to reduce anxiety and maintain composure when making decisions.
- **Team support**: Firefighters rely on strong bonds with their team members for emotional and operational support. Building a culture of trust and open

communication within firefighting teams allows individuals to express their concerns, share their experiences, and support each other through difficult situations.

- **Stress debriefing**: After particularly intense or traumatic incidents, many fire departments conduct stress debriefings to allow firefighters to process their emotions and share their experiences in a supportive environment. This practice helps reduce long-term psychological impacts and provides a space for reflection and emotional recovery.

3. Psychological Impact on Victims

Victims of emergencies may experience a wide range of emotions, including fear, confusion, helplessness, and anger. Firefighters must understand how to manage these emotional responses to ensure the safety and well-being of those they are rescuing. Some key strategies for managing victims' psychological responses include:

- **Calm communication**: Firefighters must provide clear, calm, and concise instructions to victims. Speaking in a calm tone helps reassure victims, even in the most intense situations. This can also help to reduce the likelihood of panic or irrational behavior.
- **Empathy and reassurance**: Expressing empathy and offering reassurance can help victims feel supported during an emergency. Even simple phrases like “You’re going to be okay” or “We’re here to help you” can make a significant difference in how a victim copes with the situation.
- **Active listening**: Firefighters should listen to victims’ concerns, fears, and questions, providing them with information or guidance where appropriate. This helps build trust and can improve the victim's cooperation during the rescue.

4. Building Psychological Resilience in Firefighters

Building psychological resilience is an essential part of firefighter training. This involves preparing firefighters mentally and emotionally for the challenges they will face in the field. Training programs that focus on stress management, emotional intelligence, and communication skills can significantly enhance a firefighter's ability to perform under pressure.

Key elements of building resilience include:

- **Mental toughness training**: Firefighters can be trained to develop a mental "toolbox" of strategies for coping with stress, maintaining focus, and staying positive during challenging operations.
- **Emotional intelligence**: Understanding and managing one's own emotions, as well as recognizing and responding to the emotions of others, is a crucial skill for firefighters. Emotional intelligence enables firefighters to remain level-headed during emergencies and effectively support their colleagues and the public.
- **Crisis communication skills**: Effective communication is not only about providing instructions but also about conveying reassurance and control. Firefighters who develop strong crisis communication skills can better manage both their own team and the victims they are rescuing.

Chapter 24: Stress Management Techniques

Managing stress effectively is essential for both aspiring and current firefighters, given the high-stakes nature of their work. Stress management techniques can significantly improve performance during exams and in the field, enhancing mental and physical health.

1. **Mindfulness**: Practicing mindfulness involves being fully present in the moment without judgment. This can be cultivated through meditation, deep breathing exercises, and yoga, which help reduce anxiety and improve focus. These practices are particularly beneficial before exams or challenging situations on the job, as they help calm the nervous system and clear the mind for better decision-making.
2. **Regular Physical Activity**: Engaging in physical activities like running, swimming, or weight training releases endorphins—natural mood enhancers. Establishing a consistent exercise routine not only maintains physical readiness but also provides a healthy outlet for job-related stress. Tailoring routines to individual preferences ensures sustainability over time.
3. **Adequate Sleep**: Sleep plays a crucial role in cognitive function, emotional regulation, and overall health. Firefighters should prioritize quality sleep, especially before exams or after demanding shifts. Effective sleep strategies include keeping a regular schedule, creating a restful environment, and avoiding caffeine and electronics before bed. Good sleep hygiene helps with stress management and enhances recovery.
4. **Balanced Nutrition**: A nutritious diet stabilizes mood, boosts energy, and supports health. Firefighters should focus on whole foods—such as fruits, vegetables, lean proteins, and whole grains—while limiting processed foods, sugar, and caffeine. Staying hydrated is equally important, as dehydration can affect mood and mental clarity.

5. **Strong Support System**: Building a support network of family, friends, colleagues, and mentors is invaluable. Sharing experiences and challenges with trusted individuals provides relief, perspective, and effective coping strategies. Many fire departments offer peer support programs and access to professional resources for added support.
6. **Time Management Skills**: Effective time management helps firefighters balance professional responsibilities with personal commitments and study time. Strategies include setting priorities, breaking tasks into manageable steps, and scheduling time for studying, training, and relaxation. Organized time management reduces feelings of overwhelm and boosts productivity.
7. **Positive Mindset**: Developing a positive outlook involves practicing gratitude, focusing on solutions rather than problems, and maintaining a sense of humor. A positive attitude enables firefighters to face challenges with resilience, turning stressors into opportunities for growth.
8. **Relaxation Techniques**: Techniques like progressive muscle relaxation, guided imagery, and biofeedback are effective in reducing stress and enhancing emotional well-being. Progressive muscle relaxation involves tensing and releasing muscle groups, while guided imagery encourages focusing on calming scenes. Biofeedback uses electronic monitoring to help individuals control bodily functions, such as heart rate and muscle tension.
9. **Setting realistic goals.** Firefighters face immense pressure to perform well, but setting unattainable goals can lead to frustration. By focusing on achievable, incremental goals, firefighters experience a sense of accomplishment, which reduces stress. Celebrating small victories is crucial for morale and maintaining a positive outlook.
10. **Engaging in hobbies and interests** outside of firefighting, such as reading, painting, or hiking, provides a healthy escape from job pressures. Personal interests offer mental respite and remind firefighters of life beyond the firehouse, promoting a well-rounded lifestyle.
11. **Mindfulness and reflection**, including journaling, can help firefighters process experiences constructively. Journaling helps identify stress triggers and develop effective coping strategies.

12. Seek assistance from mental health professionals if you experience overwhelming stress, anxiety, or PTSD symptoms. Professional counselors and therapists offer tailored support and strategies to navigate the psychological challenges of firefighting.

Chapter 25: Teamwork Dynamics in Firefighting

Firefighting is a team-oriented profession where success often depends on the strength of interpersonal relationships and the ability to work cohesively under extreme pressure. The dynamics of teamwork in firefighting involve not just technical skills but also a deep understanding of human relations. Effective communication, emotional intelligence, conflict resolution, and leadership are components that help firefighters navigate the challenges of high-stress situations.
This chapter delves into the essential interpersonal skills that firefighters must develop to enhance teamwork and communication in the field, ensuring operations run smoothly and efficiently, even in the most demanding environments.

1. The Importance of Effective Communication

Communication is the backbone of successful firefighting operations. In high-pressure situations, where every second counts, firefighters must be able to communicate clearly and concisely. Miscommunication or unclear instructions can lead to dangerous situations, including misallocation of resources or delayed rescue efforts.

Key aspects of communication in firefighting include:

- **Clear and concise directives**: Firefighters need to give and receive clear orders, especially in rapidly evolving situations. Brief, direct communication reduces confusion and ensures everyone understands their role.
- **Non-verbal communication**: In loud or chaotic environments where verbal communication is difficult, non-verbal signals such as hand gestures or radio signals become essential. Firefighters must be trained to recognize and use non-verbal cues effectively.

- **Listening and feedback**: Communication is not only about giving orders but also about actively listening to others. Team members must be open to feedback, whether it's from their peers, superiors, or the victims they are assisting.

2. Emotional Intelligence in Firefighting

Emotional intelligence (EI) refers to the ability to recognize, understand, and manage one's own emotions, as well as the emotions of others. In the context of firefighting, emotional intelligence is essential for maintaining composure, building strong team relationships, and supporting others during high-stress incidents.

The key components of emotional intelligence include:

- **Self-awareness**: Firefighters must be aware of their own emotional triggers and stress responses. Understanding how stress impacts decision-making can help prevent emotional reactions from impairing judgment during stressful moments.
- **Self-regulation**: Being able to control impulsive behaviors, manage stress, and stay calm under pressure is crucial for firefighters, especially in life-threatening scenarios where clear-headed thinking is required.
- **Empathy**: Firefighters regularly deal with people in distress. Empathy allows firefighters to connect with victims and team members, offering support and understanding, which improves cooperation and trust during operations.

3. Conflict Resolution and Team Dynamics

Firefighting is a team effort, and like any group dynamic, conflicts may arise. Whether it's a disagreement over strategy or a clash of personalities, firefighters must be equipped with the skills to resolve conflicts quickly and effectively to prevent disruptions in team cohesion.

Key conflict resolution strategies include:

- **Addressing issues early**: Minor disagreements can escalate if not dealt with promptly. Firefighters should address concerns as soon as they arise, focusing on finding solutions rather than assigning blame.
- **Collaborative problem-solving**: Encouraging open dialogue and involving all parties in the decision-making process fosters a sense of teamwork and shared responsibility. This approach helps resolve conflicts by ensuring everyone feels heard and valued.
- **Maintaining professionalism**: Even in the heat of the moment, firefighters must remain professional. Personal conflicts should never interfere with the operational goals of the team.

By fostering a culture of respect and open communication, teams can resolve conflicts efficiently and maintain strong, effective working relationships.

4. Cultural Competence in Firefighting

Firefighters work in diverse communities and alongside team members from various backgrounds. Cultural competence—the ability to interact respectfully and effectively with people from different cultural, racial, and socioeconomic backgrounds—is increasingly important in modern firefighting.

Building cultural competence involves:

- **Recognizing diversity**: Firefighters must acknowledge and respect the different cultural practices, values, and communication styles that exist within their team and the communities they serve.
- **Practicing inclusivity**: Inclusivity fosters an environment where all team members feel valued and respected, regardless of their background. A diverse team brings varied perspectives and approaches, enriching the overall effectiveness of firefighting efforts.
- **Understanding bias**: Firefighters must be aware of their own implicit biases and work actively to ensure these biases do not affect their treatment of

colleagues or members of the public. Recognizing and overcoming unconscious biases leads to better teamwork and community relations.

5. Leadership Skills in Firefighting

Strong leadership is essential for coordinating firefighting operations, ensuring safety, and motivating teams during high-stress situations.

Key leadership traits include:

- **Decisiveness**: In emergencies, leaders must make quick, informed decisions. Indecisiveness can lead to confusion and delay actions.
- **Confidence and authority**: Firefighters look to their leaders for direction. Leaders must project confidence and take responsibility for their decisions to maintain team trust.
- **Team empowerment**: Effective leaders recognize the strengths of their team members and delegate responsibilities accordingly. By empowering team members to take initiative, leaders create a more capable and resilient team.
- **Mentorship and support**: Firefighting leaders serve as mentors, guiding less experienced firefighters and providing ongoing support to all team members. Good leaders foster an environment of learning and personal growth within the team.

Emotional Responses in Emergencies

Managing emotional responses in emergency situations requires firefighters to develop a deep understanding of their own emotions and those of others.

One of the first steps in managing emotional responses is recognizing the signs of stress and anxiety, both in oneself and in others. These signs can include changes in breathing, increased heart rate, confusion, or inability to focus. By identifying these signs early, firefighters can employ strategies to mitigate their impact.

Breathing exercises are a powerful tool for managing emotional responses. Deep, controlled breathing can help reduce stress levels, lower heart rate, and clear the mind, allowing for better decision-making and performance under pressure. Encouraging victims and team members to engage in breathing exercises can also help them regain a sense of calm in chaotic situations.

Positive self-talk is another effective strategy for managing emotions. This involves consciously shifting negative or fearful thoughts to positive affirmations or realistic assessments of the situation. For example, instead of thinking, "We can't handle this," a firefighter might remind themselves, "We've trained for situations like this. We can do it."

Peer support Firefighters should be encouraged to lean on each other for emotional support, sharing their feelings and experiences openly. This not only helps individuals process their emotions but also strengthens the bond between team members, creating a supportive network that can navigate challenges more effectively.

Training and preparation are key to managing emotional responses. Regular drills and simulations that mimic emergency situations can help firefighters become accustomed to the stress and chaos they might face, reducing the intensity of their emotional reactions when a real emergency occurs. This training should also include components that focus on emotional resilience, such as scenario-based discussions about managing fear and anxiety.

Debriefing after incidents is crucial for processing emotional experiences. These sessions provide an opportunity for firefighters to discuss what happened, how they felt, and how they managed those feelings. Debriefings can be facilitated by mental health professionals when necessary, offering an additional layer of support and strategies for coping with difficult emotions.

Mindfulness and meditation practices can enhance emotional regulation over time. By cultivating a practice of mindfulness, firefighters can improve their ability to remain present and focused, even in the face of fear or stress. Meditation can also reduce overall stress levels, making it easier to manage emotional responses in the moment.

Physical fitness is not only essential for the physical demands of firefighting but also for emotional well-being. Regular exercise can help reduce stress, improve mood, and increase resilience to emotional challenges. Encouraging a culture of physical wellness within the fire department can have a positive impact on emotional regulation.

Educational resources about emotional intelligence and stress management should be made readily available to firefighters. Workshops, seminars, and access to online materials can provide valuable information and tools for understanding and managing emotions effectively.

Communication and Coordination Techniques

Clear, concise, and direct communication ensures that all team members are aware of their roles, responsibilities, and the current situation, which is essential for the safety and efficiency of operations. To facilitate this, **standardized communication protocols** should be established and practiced regularly. These protocols include the use of common terminology, clear radio communication practices, and predefined signals and gestures that can be understood even in noisy or visually obstructed environments.

Regular training and drills that involve multiple agencies can significantly improve coordination and communication during real emergencies. These exercises should simulate real-life scenarios as closely as possible, allowing firefighters to practice working with law enforcement, medical personnel, and other relevant agencies. This not only helps in understanding the capabilities and limitations of each group but also builds mutual respect and trust, which are invaluable during actual incidents.

Debriefing sessions after training exercises and emergency responses offer another opportunity to enhance communication and coordination. These sessions provide a platform for open discussion about what worked well and what did not, allowing teams to make necessary adjustments to their strategies and communication methods. It is important that these debriefings are conducted in a constructive manner, with a focus on learning and improvement rather than assigning blame.

Technology also plays a crucial role in improving communication and coordination. The use of mobile data terminals (MDTs), smartphones, and other communication devices can provide real-time information sharing and updates on the situation, resource availability, and other factors. However, it is essential to ensure that all team members are trained in the use of these technologies and that backup communication methods are in place in case of technology failure.

Incident command systems (ICS) An ICS provides a standardized approach to the command, control, and coordination of emergency response, offering a clear chain of command and a framework for managing emergency operations. Familiarity with the ICS structure and roles can greatly enhance the efficiency of inter-agency operations.

Listening skills are as important as speaking or issuing commands. Effective communication is a two-way street, requiring attention to feedback and input from all team members. Active listening can help identify potential issues before they escalate, as well as foster a culture of inclusivity and respect within the team.

Emotional intelligence is an often-overlooked aspect of communication. Understanding and managing one's own emotions, as well as recognizing and responding appropriately to the emotions of others, can prevent misunderstandings and conflicts. Firefighters with high emotional intelligence can effectively navigate the stress and chaos of emergency situations, maintaining clear and calm communication even under pressure.

Composure and Decision-Making Strategies

Developing these abilities requires a combination of mental preparation, practical experience, and continuous learning. **Mental preparation** involves engaging in activities that simulate the stress of emergency situations, such as high-intensity training exercises or mental visualization techniques. These activities help condition the mind to remain calm and focused when confronted with real-world emergencies. **Practical experience** is gained through participation in actual firefighting operations and drills, where the theoretical knowledge is applied in practice. This hands-on experience is invaluable for understanding how to navigate complex situations and make quick, effective decisions.

Continuous learning is essential for staying updated on the latest firefighting techniques and strategies. This includes attending workshops, seminars, and courses that focus on stress management, decision-making processes, and emergency response tactics. By embracing a mindset of lifelong learning, firefighters can continuously improve their ability to maintain composure and make sound decisions under stress.

Scenario analysis is another effective strategy for enhancing decision-making skills. This involves reviewing past emergency situations, either real or simulated, and discussing alternative actions and outcomes. By analyzing different scenarios, firefighters can better anticipate potential challenges and develop strategies for addressing them efficiently.

Peer feedback plays a crucial role in refining decision-making abilities. Constructive criticism from colleagues and superiors helps identify areas for improvement and encourages personal and professional growth. Encouraging an environment where feedback is openly shared and received can significantly enhance a firefighter's ability to make sound decisions under pressure.

Stress reduction techniques, such as deep breathing, meditation, and physical exercise, are also beneficial for maintaining composure. Incorporating these practices into daily routines can help lower stress levels, improve focus, and enhance overall well-being, which are crucial for effective decision-making in emergency situations.

Team debriefings after emergency responses offer valuable learning opportunities. These sessions allow team members to discuss the decisions made during an incident, evaluate their effectiveness, and identify lessons learned. Debriefings can reinforce positive outcomes and provide insights into how decision-making processes can be improved in future responses.

Resilience training is designed to strengthen the mental and emotional fortitude needed to cope with the pressures of firefighting. This training can include exercises aimed at building confidence, fostering a positive attitude, and developing coping strategies for managing stress and adversity.

Simulation training using virtual reality (VR) or other immersive technologies can replicate the conditions of real-life firefighting scenarios in a controlled environment. This allows firefighters to practice maintaining composure and making decisions under stress without the risks associated with actual emergencies.

Mindfulness and situational awareness are critical for effective decision-making. Being fully present and aware of the environment and its challenges enables firefighters to assess situations accurately and respond appropriately. Training in mindfulness and situational awareness can help improve concentration, reduce impulsivity, and enhance decision-making skills.

Cultural Competence in Emergency Teams

Cultural competence in the context of firefighting is not just about recognizing and respecting diversity but also about leveraging this understanding to enhance teamwork and effectiveness in emergency situations. Firefighters often encounter individuals from various cultural backgrounds, each with its unique perspectives, beliefs, and practices, especially during high-pressure scenarios. Recognizing these differences and knowing how to communicate effectively across cultures can significantly impact the success of firefighting operations and the safety of both firefighters and the communities they serve.

Effective Communication is crucial. Misunderstandings or misinterpretations can lead to delays or errors in emergency response. Firefighters must be adept at adjusting their communication styles to suit the cultural context of the individuals they are assisting. This might involve being aware of non-verbal cues, understanding the importance of certain cultural practices, or even knowing basic phrases in another language to facilitate initial communication.

Team Dynamics benefit greatly from cultural competence. Firefighting teams are increasingly diverse, reflecting the multicultural nature of the communities they protect. A culturally competent team can leverage the unique insights and perspectives of its members to solve problems more creatively and effectively. Understanding and respecting each team member's cultural background fosters a more inclusive and supportive work environment, which is essential for maintaining morale and teamwork during stressful situations.

Community Engagement strategies improve when firefighters possess cultural competence. Engaging with the community in non-emergency settings helps build trust and understanding. Firefighters who are familiar with the cultural norms and values of the communities they serve can tailor safety education and prevention programs to be more effective and well-received. This proactive approach not only helps in preventing emergencies but also ensures smoother operations when emergencies do occur.

Decision Making in emergency situations can be enhanced by cultural knowledge. Certain cultural or religious practices might influence individuals' behavior during emergencies, such as reluctance to leave a place without religious artifacts or hesitation to receive aid from someone of a different gender. Understanding these nuances allows firefighters to make informed decisions that respect cultural sensitivities without compromising safety.

Training and Development programs within fire departments should include modules on cultural competence. This training can cover a wide range of topics, from basic cultural awareness to more in-depth studies of specific cultural practices and communication styles. Regular, ongoing training ensures that firefighters are well-prepared to interact effectively with diverse populations.

Policy Development should also reflect an understanding of cultural competence. Policies and procedures that acknowledge and address the needs of diverse communities can guide firefighters in providing equitable and effective services. This might include protocols for working with interpreters, guidelines for engaging with cultural leaders in the community, or standards for culturally sensitive communication.

Diversity Training in Firefighting Teams

Building on the foundation of cultural competence, diversity training in firefighting teams extends into practical strategies and exercises that enhance team cohesion and operational efficiency. Effective diversity training programs are designed to address and bridge gaps in understanding and collaboration among team members from varied backgrounds. These programs often include interactive workshops, scenario-based learning, and discussions that encourage open dialogue about diversity and inclusion.

One key aspect of diversity training involves role-playing exercises that simulate emergency situations with diverse cultural contexts. These exercises allow firefighters to practice and refine their communication skills, adaptability, and problem-solving strategies in a controlled, reflective environment. By engaging in these role-plays, firefighters can better understand the perspectives of their colleagues and the communities they serve, leading to more empathetic and effective responses in real-world scenarios.

Implicit bias refers to the attitudes or stereotypes that affect our understanding, actions, and decisions in an unconscious manner. Recognizing and mitigating implicit biases can significantly improve team dynamics and decision-making processes. Workshops focused on this area help firefighters identify their own biases and learn strategies to minimize their influence on their work.

Diversity training also covers conflict resolution techniques that are essential for maintaining harmony within diverse teams. Firefighting is a high-stress profession where conflicts, if not managed properly, can undermine team effectiveness and safety. Training sessions that focus on conflict resolution equip firefighters with the skills to navigate disagreements constructively, ensuring that differences in opinion or approach strengthen rather than weaken the team's resolve.

In addition to these components, diversity training emphasizes the importance of emotional intelligence in fostering a supportive and inclusive team environment. Emotional intelligence—the ability to understand and manage one's own emotions, as well as recognize and influence the emotions of others—plays a crucial role in building strong, cohesive teams. Through activities that enhance self-awareness, self-regulation, motivation, empathy, and social skills, firefighters can improve their interactions with colleagues and community members alike.

Finally, diversity training often includes education on the historical and social contexts that shape the experiences of different cultural groups. This broader understanding helps firefighters appreciate the significance of cultural competence and the impact of their actions on community relations. By grounding diversity training in the realities of societal diversity and inequality, fire departments can cultivate a workforce that is not only skilled in firefighting techniques but also adept at navigating the complexities of the modern, multicultural world.

Workplace Behavior in Firefighting Teams

Creating a positive and inclusive workplace within firefighting teams requires a proactive approach to addressing bias, fostering respect, and enhancing collaboration. Bias, whether implicit or explicit, can significantly impact team dynamics, decision-making, and the overall effectiveness of emergency response efforts. To combat this, fire departments must implement strategies that promote understanding and mitigate prejudicial attitudes. This involves regular training sessions focused on diversity and inclusion, which can help team members recognize and address their biases. Such training should not be a one-time event but an ongoing process that evolves with the team and the community it serves.

Respect is the cornerstone of effective teamwork, especially in high-pressure environments like firefighting. It goes beyond mere tolerance of differences to encompass a deep appreciation for the unique perspectives and skills each member brings to the team. Respectful interactions are characterized by active listening, open communication, and the validation of others' experiences and opinions. Fire departments can cultivate respect by creating forums for open dialogue, where team members can share their experiences and learn from one another in a supportive setting.

Collaboration within diverse communities and firefighting teams is essential for addressing the complex challenges of emergency response. A collaborative team leverages the diverse strengths and abilities of its members to achieve common goals. This requires a clear understanding of roles and responsibilities, as well as trust in each other's capabilities. Building such trust involves team-building exercises and shared experiences that reinforce the team's cohesion and interdependence. Moreover, collaboration extends beyond the immediate team to include other emergency response agencies and the community at large. Engaging with community leaders and members can provide valuable insights into cultural considerations and local knowledge that enhance the team's effectiveness.

To ensure these principles are not only understood but also practiced, fire departments should establish clear policies and procedures that outline expected behaviors and the consequences of failing to adhere to these standards. Regular assessments of workplace culture and team dynamics can help identify areas for improvement and measure the impact of diversity and inclusion initiatives. Additionally, leadership plays a crucial role in modeling appropriate behavior and setting the tone for the entire team. Leaders who demonstrate a commitment to diversity, respect, and collaboration can inspire their teams to follow suit, creating a positive and inclusive workplace culture that benefits everyone.

Part J: Local Knowledge

Chapter 26: Understanding Local Geography

Understanding local geography is a fundamental aspect of firefighting that cannot be overstated. Knowing the layout of streets, the types of buildings within a community, and the locations of hydrants and water sources equips firefighters with the necessary information to respond efficiently and effectively to emergencies. This knowledge allows for quicker response times, strategic planning of attack routes, and the identification of potential water supply issues before they become significant during operations.

Firefighters must be familiar with the unique characteristics of their service area, including residential, commercial, and industrial zones, as each presents its own set of challenges and risks. Residential areas, for example, may have narrow streets that could hinder the maneuverability of fire apparatus, while industrial zones could house hazardous materials that require specialized response strategies.

The importance of local geography extends beyond the physical layout of an area. Understanding the community includes knowing about buildings that are historically significant, those that are prone to specific types of incidents, or areas that are under development and might not yet appear in standard mapping resources. This depth of knowledge can only be achieved through diligent study and continuous updating of information, as communities are ever-changing. Firefighters often use a variety of tools to maintain and update their geographical knowledge, including GIS (Geographic Information Systems), detailed maps, and pre-incident planning software. These tools, combined with regular physical surveys of the area, ensure that firefighters are not caught off-guard by new developments or changes in their operational environment.

Preparation for location-specific exams requires candidates to demonstrate their understanding of local geography. These exams may include questions designed to test knowledge of street layouts, the identification of key buildings or landmarks, and scenarios that require the candidate to choose the best route or water source under time pressure. Success in these exams reflects a candidate's readiness to navigate their local area under the stress of real emergency conditions. The study of local geography for these exams should be approached methodically, starting with a broad overview of the area and gradually focusing on more detailed aspects. Candidates are encouraged to walk or drive through their areas of responsibility, taking note of potential hazards, access points, and water supply options. This hands-on approach not only aids in memorization but also helps build a spatial understanding of the area that can be crucial during an emergency response.

In addition to the practical benefits of understanding local geography, this knowledge also fosters a deeper connection between firefighters and the communities they serve. By being familiar with the area, firefighters can engage more effectively with residents, participate in community safety initiatives, and tailor fire prevention efforts to address local risks. This community-focused approach to firefighting underscores the role of firefighters not just as emergency responders but as integral members of the community with a vested interest in its safety and well-being.

For firefighter candidates, mastering the intricacies of local geography is not just about passing an exam; it's about preparing for the reality of the job. The ability to quickly navigate through diverse terrains, from densely populated urban areas to secluded rural settings, can mean the difference between life and death in emergency situations. This requires a comprehensive understanding of not only the physical layout but also the socio-economic factors that influence the nature of incidents in different areas. For instance, areas with a higher concentration of older buildings might be more susceptible to fires due to outdated electrical systems, necessitating a different approach to fire prevention and response.

The role of technology in enhancing geographical knowledge cannot be understated. Advanced mapping tools and mobile applications provide real-time data that can be crucial during emergency responses. These technologies allow firefighters to access detailed information about building structures, potential hazards, and hydrant locations at the touch of a button. However, reliance on technology should not replace the fundamental need for personal familiarity with the local geography. Regular training exercises that simulate real-life scenarios in different parts of the community are essential for reinforcing this knowledge and ensuring that firefighters can effectively apply it when it matters most.

New construction projects, road closures, and modifications to water systems can significantly impact firefighting strategies. Being proactive in obtaining this information ensures that firefighters are always one step ahead, ready to adapt their tactics to the evolving landscape of their service area.

The integration of local geography knowledge into the firefighting curriculum goes beyond practical applications; it instills a sense of pride and ownership among firefighters. Knowing the history, culture, and unique challenges of their community deepens their commitment to serving it with excellence. This holistic approach to understanding local geography not only prepares firefighter candidates for the technical aspects of their role but also for the profound responsibility of protecting their community.

As candidates prepare for their exams, they should view the study of local geography not as a mere academic exercise but as a foundational element of their future role as firefighters. Engaging with the material with curiosity and a deep sense of purpose will enrich their learning experience and set them on the path to becoming knowledgeable, effective, and compassionate firefighters. This comprehensive grasp of local geography, combined with technical skills and a commitment to community service, forms the cornerstone of exceptional firefighting, ready to meet the challenges of the modern world with confidence and expertise.

Part K: Physical Fitness

Chapter 27: Fitness for Firefighters

The physical requirements for firefighters are rigorous and multifaceted, demanding a high level of cardiovascular fitness, strength, endurance, and flexibility. Cardiovascular fitness is paramount, as firefighters often work in situations that elevate their heart rate for extended periods. This could include carrying heavy equipment up several flights of stairs, rescuing victims, or even battling a blaze. To build and maintain a robust cardiovascular system, candidates and seasoned firefighters alike should engage in a variety of aerobic exercises such as running, swimming, cycling, or rowing. These activities help improve the heart's efficiency in pumping blood and delivering oxygen to the muscles.

The nature of firefighting work often involves lifting, dragging, and hoisting heavy equipment and, potentially, victims. Incorporating strength training into a fitness regimen is essential. Exercises focusing on major muscle groups through compound movements like squats, deadlifts, bench presses, and overhead presses can provide the foundational strength needed. Additionally, functional fitness exercises that mimic real-life firefighting activities can be particularly beneficial. These might include sled drags, tire flips, and weighted stair climbs, which simulate the physical demands encountered on the job.

Endurance and flexibility are equally important for firefighters. Endurance training, which can include longer sessions of moderate-intensity cardiovascular exercises, helps improve the body's ability to sustain activity over time, crucial during extended firefighting operations.

Flexibility exercises, such as dynamic stretching before physical activity and static stretching afterward, help prevent injuries and ensure that firefighters can move freely and efficiently in the demanding and unpredictable environments they often face.

The **PAT (Physical Ability Test)** and **CPAT (Candidate Physical Ability Test)** are both physical fitness tests designed to assess the physical readiness of firefighter candidates. Here's a breakdown of each:

1. **PAT (Physical Ability Test)**: This is a general term for physical assessments used by many fire departments to evaluate candidates' strength, stamina, and ability to perform essential firefighting tasks. The PAT may vary by department but typically includes tasks such as carrying heavy equipment, climbing stairs, dragging hoses, and simulating rescues. These tests are tailored to replicate the physical challenges firefighters face on the job.
2. **CPAT (Candidate Physical Ability Test)**: This is a standardized, widely recognized test used by many fire departments across North America. The CPAT is designed to be a fair and consistent measure of the physical capabilities required for firefighting. It consists of eight specific tasks:
    - **Stair Climb**: Climbing stairs with a weighted vest to simulate carrying equipment in high-rise buildings.
    - **Hose Drag**: Dragging a hose to demonstrate lower body strength and coordination.
    - **Equipment Carry**: Carrying tools over a set distance, testing grip and upper body strength.
    - **Ladder Raise and Extension**: Raising and extending a ladder to test upper body strength and coordination.
    - **Forcible Entry**: Using a sledgehammer to strike a target, simulating breaking through barriers.
    - **Search**: Crawling through a dark maze, testing spatial orientation and stamina.
    - **Rescue**: Dragging a weighted mannequin, simulating a rescue scenario.

- **Ceiling Breach and Pull**: Using a pike pole to push and pull a ceiling, testing upper body strength and endurance.

The **CPAT** is often preferred by fire departments as it provides a consistent, validated assessment of a candidate's fitness and job readiness, ensuring they have the physical capabilities to handle the demands of firefighting.

Preparing for the CPAT, a specific physical test for firefighters, requires a targeted approach. This test includes eight challenging events designed to simulate the physical tasks associated with firefighting. To prepare effectively, candidates should familiarize themselves with the CPAT events and structure their training to build the necessary skills and physical capabilities. This might involve practicing ladder raises, hose drags, and equipment carries, among other activities. It's also beneficial to incorporate interval training into workouts to mimic the high-intensity bursts of activity followed by periods of lower intensity or rest, reflecting the real-world demands of firefighting.

Beyond the CPAT, maintaining overall health and wellness is crucial for firefighters. This encompasses a balanced diet rich in nutrients, adequate hydration, sufficient sleep, and stress management techniques. A healthy lifestyle supports the physical demands of the job and aids in recovery, resilience, and long-term well-being. Firefighters should also be proactive in seeking regular medical check-ups to monitor their health status and address any issues early. The psychological demands of firefighting can be significant, with exposure to traumatic events, high-stress situations, and the physical toll of the job contributing to mental health challenges. Strategies for mental health maintenance might include mindfulness practices, counseling, and peer support groups, ensuring firefighters are supported both physically and mentally.

In preparing for the physical demands of firefighting, it's essential to adopt a comprehensive and consistent training regimen that addresses cardiovascular fitness, strength, endurance, and flexibility. This approach, combined with a focus on overall health and wellness, equips firefighter candidates and seasoned professionals alike to meet the challenges of the job head-on. As we delve further into the specifics of each physical requirement and the best practices for training and maintenance, it becomes clear that a holistic approach to physical fitness and health is not just beneficial but necessary for the demanding and rewarding career of firefighting.

Building on the foundation of cardiovascular fitness, strength, endurance, and flexibility, it's important to highlight the role of nutrition and hydration in a firefighter's physical preparedness and overall health. A diet that emphasizes lean proteins, whole grains, fruits, vegetables, and healthy fats provides the energy and nutrients necessary for peak performance and recovery. Firefighters should aim to drink water consistently throughout the day, not just during or after physical exertion, to maintain optimal hydration levels.

Sleep is another cornerstone of physical and mental health that cannot be overlooked. Quality sleep supports recovery from physical exertion, helps consolidate memory, and regulates mood. Firefighters, who often work irregular hours and may be exposed to high levels of stress, should prioritize sleep to maintain their health and performance. Strategies to improve sleep include establishing a regular sleep schedule, creating a restful environment free of distractions, and practicing relaxation techniques before bed.

Chronic stress can lead to a range of physical and mental health issues, including cardiovascular disease, anxiety, depression, and burnout. Effective stress management techniques include regular physical activity, which itself is a powerful stress reliever, as well as mindfulness meditation, deep breathing exercises, and engaging in hobbies or activities that provide relaxation and enjoyment.

In addition to individual efforts to maintain physical fitness and health, fire departments can play a supportive role by providing resources and programs that promote the well-being of their personnel. This might include access to fitness facilities, nutritional counseling, stress management workshops, and mental health services. Creating a culture that values and supports health and wellness can encourage firefighters to take proactive steps toward maintaining their fitness and well-being.

Finally, it's essential for firefighters to recognize the signs of overtraining and injury, which can result from an overly aggressive approach to physical preparation. Symptoms of overtraining include excessive fatigue, decreased performance, persistent soreness, mood disturbances, and increased susceptibility to illness. Listening to one's body and allowing adequate time for rest and recovery is crucial in preventing overtraining and ensuring long-term health and fitness.

To claim your bonuses, please send an email to Zenith24Editions@gmail.com with the subject line "Firefighter Exam Prep Bonuses". You will receive your 1000+ Q&A Workbook, the Flashcards and the audiobook for free.

Conclusion

Becoming a firefighter is not just about physical strength or technical knowledge; it requires a combination of mental resilience, strong interpersonal skills, and a deep commitment to serving the community. The *Firefighter Exam Prep* book has provided you with the foundational knowledge and practical skills necessary to excel in both the firefighter entrance exam and the day-to-day challenges you will face in the field.

From understanding fire science and equipment operation to mastering mechanical reasoning and human relations, this guide has covered the wide range of topics essential for firefighting success. Throughout this journey, you have explored not only the tactical aspects of firefighting but also the psychological and emotional resilience needed to thrive in high-pressure situations.

Remember, the path to becoming a firefighter is a lifelong learning process. As technologies evolve and new challenges emerge, continuous education and training are key to maintaining your effectiveness in the field. Moreover, your role as a firefighter will demand ongoing growth—not just in technical proficiency, but in your ability to connect with and support your team and the community you serve.

As you move forward, keep in mind the core values that define firefighting: bravery, integrity, teamwork, and dedication. You are entering a profession that requires not only skill but also an unwavering commitment to the safety and well-being of others. By applying the knowledge and strategies you've learned in this book, you will be well-prepared to face the challenges ahead, protect lives, and contribute to the legacy of service that defines the firefighting community.

We wish you the best of luck on your exam and in your future career as a firefighter. Stay committed, stay focused, and remember that the impact you will have as a firefighter goes beyond the fires you extinguish—it touches the lives of everyone in the community you serve.

Made in the USA
Las Vegas, NV
11 December 2024